ELECTRONIC DESIGN

WITH OFF-THE-SHELF

INTEGRATED CIRCUITS

ELECTRONIC DESIGN
WITH OFF-THE-SHELF
INTEGRATED CIRCUITS

Z. H. Meiksin

and

Philip C. Thackray

PARKER PUBLISHING COMPANY, INC.

WEST NYACK, NEW YORK

© 1980 *by*

Parker Publishing Company, Inc.
West Nyack, New York

Library of Congress Cataloging in Publication Data

Meiksin, Z   H
  Electronic design with off-the-shelf integrated
circuits.

  Includes index.
  1. Integrated circuits. 2. Electronic
circuit design. 3. Electronic apparatus and
appliances--Design and construction. I. Thackray,
Philip,   joint author. II. Title.
TK7874.M38   621.381'73   79-12389
ISBN 0-13-250282-8

Printed in the United States of America

Dedication

To my wife Jeannine—
    Z.H. Meiksin

To my parents—
    Philip C. Thackray

Acknowledgment

The diligent work of Beverly Bocheff in typing and proofreading portions of the manuscript is gratefully acknowledged.

A WORD FROM THE AUTHORS

ON THIS BOOK'S PRACTICAL VALUE

This book shows how to select the appropriate analog and digital integrated circuits (I.C.'s) and the peripheral passive components and how to design these components into functional circuits and systems. The emphasis is on functional design rather than analysis. Practical considerations such as bias currents, offset voltages, drift and noise are incorporated in the design procedures and precise examples are given.

Vast advances made by the electronic industry in the manufacture and supply of standard I.C.'s have a direct effect on the design of electronic systems and circuits. You can now build considerably more complex systems with less effort and at a lower cost than ever before, while thinking in terms of sub-systems rather than individual components. This book concentrates on *practical* approaches. It shows, for example, how to choose just the right amplifier for an oscillator circuit with specified characteristics; and it provides tables for the design of active filters that must meet given specifications. Numerical examples and potential pitfalls are given throughout the book.

I.C.'s are considered building blocks. The operational amplifier, which is the fundamental block of analog systems, is introduced in Chapter 1; the basic gates and building blocks of digital systems are introduced in Chapters 8 and 9. Numerous applications are given in these and other chapters of the book. The chapters can be read independently allowing you to turn directly to a subject of immediate interest by consulting the Table of Contents and the extensive Index.

Chapter 2 discusses passive components including resistors, capacitors, inductors, and transformers. It explains why, for example, a carbon resistor must be chosen in one case and a wire wound resistor in another. Since the internally generated noise in active and passive components is ever present, special consideration must be given to noise effects in low noise circuits. How to design low-noise circuits is shown in

Chapter 3. Chapter 4 treats sinusoidal and square wave generators for analog and digital systems. Linear applications are presented in Chapter 5 which includes a variety of amplifiers, integrators, current-to-voltage converters and many other circuits. Nonlinear applications are given in Chapter 6. This chapter includes precision rectifier circuits using operational amplifiers, peak detectors, sample-hold circuits, comparators, level detectors, and zero-crossing detectors with several variations. Designs of active filters are given in Chapter 7. Step-by-step design procedures for low-pass, high-pass, and band-pass filters are given in an easy to follow manner with the aid of tables.

Basic principles of digital circuits and simple digital logic I.C.'s are introduced in Chapter 8, while MSI and LSI I.C.'s and their usage are given in Chapter 9. Practical ADA conversion schemes are presented in Chapter 10.

Important features of the book are the topics of grounding and shielding, and of *complete* system design. These topics are covered in Chapters 11 and 12, respectively. The topic of grounding and shielding is presented by including the realistic, practical steps necessary to insure interference-free operation. Since the techniques applicable to analog systems are often incompatible with those applicable to digital systems, the techniques applicable to each case are described separately. It is then shown how to deal with systems which incorporate both analog and digital circuits. The chapter on system design presents by means of specific examples, and in step-by-step-detail, how to proceed in system design from a basic concept, through the block diagram, to the final circuit with the proper selection of components. The great majority of the many circuits in this book have actually been assembled and field tested. Thus, accuracy and reliability have been confirmed.

Z. H. Meiksin
Philip C. Thackray

CONTENTS

THE OPERATIONAL AMPLIFIER *(continued)*

CONTENTS .

DESIGNING LOW NOISE CIRCUITS *(continued)*

CONTENTS

CONTENTS

ELECTRONIC DESIGN

WITH OFF-THE-SHELF

INTEGRATED CIRCUITS

Chapter 1

The Operational Amplifier

1.1   The central element of modern linear electronic circuits is the operational amplifier. We distinguish between *voltage operational amplifiers,* and *transconductance operational amplifiers.* A *voltage* operational amplifier is a high voltage gain, high input impedance, low output impedance amplifier. The term "operational" is used for historical reasons, as this type of amplifier was first developed to perform mathematical operations in analog computers. A *transconductance* operational amplifier is a high input impedance, high output impedance amplifier characterized by its linear transconductance, that is, by an output current proportional to the input voltage. The voltage operational amplifier is by far the most commonly used amplifier. In this book the term "operational amplifier" will imply *voltage* operational amplifier, unless qualified by the adjective "transconductance." The operational amplifier underwent several states in its development in rapid succession: vacuum tube amplifier, discrete transistor amplifier, and integrated circuit (I.C.) amplifier. The I.C. amplifier achieved a high degree of perfection, and a large variety of amplifiers is available.

The objective of this chapter is to show how to use operational amplifiers in basic important applications. Many more specialized applications are introduced in later chapters. Real amplifiers have non-ideal limitations which introduce inaccuracies. The main thrust of this chapter is to show how to take these limitations into consideration in the design of practical circuits. The consideration of the characteristics of *real* amplifiers and *real* components in designs is the main purpose of the book.

1.2   THE BASIC AMPLIFIER

The symbol of an operational amplifier is shown in Figure 1-1. All the voltages are referenced to ground as shown in Figure 1-2. The voltages V^+ and V^- are the positive and negative power supplies also referenced to ground. In most applications the operational amplifier is connected to positive and negative power supplies. The amplifier has no ground terminal, but the voltages in the amplifier circuit become referenced to ground

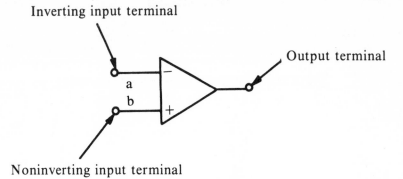

Figure 1-1. The operational amplifier symbol.

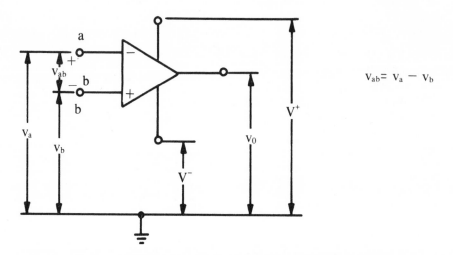

Figure 1-2. The operational amplifier voltages referenced to ground.

v_a = inverting input voltage
v_b = noninverting input voltage
v_o = op amp output voltage
V^+ = positive voltage of power supply
V^- = negative voltage of power supply

through the power supplies and through the external input and output circuits. There are also circuits in which the operational amplifier operates from a single polarity power supply. In this case one terminal of the amplifier is connected to ground.

The operational amplifier can be utilized with three different configurations of the external circuitry. The signal voltage can be applied to the *inverting terminal* input (terminal a in Figure 1-2), to the *noninverting terminal* (terminal b), or as a differential input, between terminals a and b. Each configuration offers unique characteristics and applications. The manufacturer supplies the application engineer with a set of specifications for each operational amplifier type. These specifications impose constraints on the design and are ultimately responsible for particular choices of components. We shall explain the concepts and terms used in the specifications, and in the last section of this chapter we show, by example, how these specifications must be included in the design procedure of amplifiers.

1.2.1 Gain

The gain A of an operational amplifier is specified in terms of the *differential* input voltage v_{ab} and the output voltage.

$$A = - \frac{v_o}{v_{ab}}$$

The output voltage v_o is measured between the output terminal and ground. The negative sign in the equation indicates that if the voltage at terminal a is positive with respect to the voltage at terminal b, the voltage at the output terminal is *negative* with respect to ground. The gain A is a function of the input voltage frequency f: the gain is highest for d.c. and low frequency signals, and decreases with increasing frequency. For stable operation the *gain versus frequency* characteristic must meet certain specifications, depending upon the external circuitry in which the amplifier is used. To meet these characteristics, certain operational amplifiers are *compensated* internally by the manufacturer, while other operational amplifiers must be compensated by the user. Each type has a field of application. Section 1.4 deals in detail with the compensation problem.

The gain-frequency response curves of an internally compensated operational amplifier (type 741), and of an uncompensated operation amplifier (type 748) are shown respectively in Figure 1-3 and Figure 1-4. Note the difference in the magnitudes of the frequencies at the respective 3 dB points, and the differences in the slopes beyond the 3 dB points.

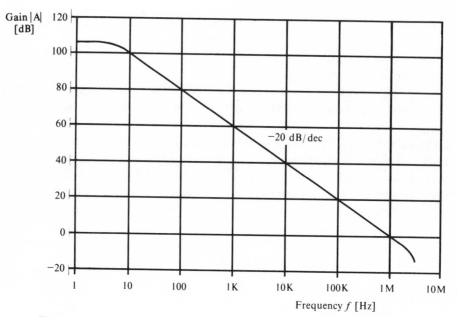

Figure 1-3. Gain-frequency response curve of operational amplifier type 741, an internally compensated amplifier. Note the low frequency of about 5 Hz at the "3 dB point" and the "bandwidth" of 1 MHz (i.e. frequency at unity gain or 0 dB gain).

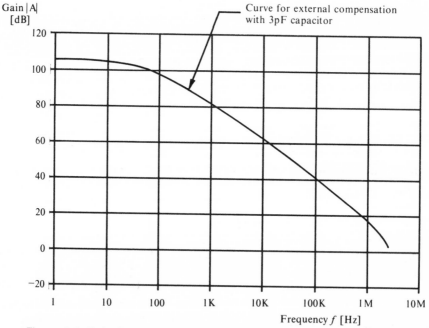

Figure 1-4. Gain-frequency response curve of operational amplifier type 748 with external compensation. Note the higher frequency of about 50 Hz at the "3dB point" and the higher frequencies for comparable gains of the 741. For example, at 40dB the 748 has a frequency response of 100 KHz, compared to 10 kHz of the 741.

Operational amplifiers are characterized by high d.c. gain values on the order of hundreds of thousands. The gain is often specified in decibels (dB). The decibel is defined as 20 times the logarithm (to the base 10) of a number:

$$\text{Gain (dB)} \ = \ 20 \log_{10} A$$

For example, a gain of 100,000 corresponds to a gain of 100 dB.

$$20 \log_{10} 100,000 \ = \ 20 \times 5 \ = \ 100 \text{ dB}$$

1.2.2   Input Resistance

The input resistance of an amplifier is defined as the resistance between the inverting and the noninverting input terminals. It is shown symbolically as R_i in Figure 1-5. Values lie in the range from hundreds of kilo-ohms to hundreds of megohms. The input of the 741 amplifier, for example, is specified as *"typical"* $1\,\text{M}\Omega$ and *minimum* $300\text{K}\Omega$. The LM 741 has a junction transistor input stage. Tens and hundreds of megohms input resistances are available in operational amplifiers with field effect transistor input stages (e.g. the LH0042), or "super beta" transistor input stages (e.g. the LM108). The input capacitance is very small, on the order of a few picofarads.

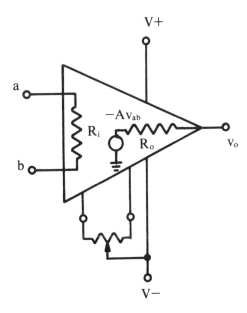

Figure 1-5. Voltage offset balancing circuit. (Also shown symbolically are input and output resistances.)

1.2.3 Output Resistance

The output resistance of an operational amplifier constitutes circuitry responsible for the difference between the amplifier output voltage under no-load and under load, due to current flowing in this part of the circuit under load. It is represented symbolically as R_o in Figure 1-5. It can be seen from the figure that if no load is connected to the output terminal of the amplifier, no current flows through R_o and the output voltage is equal to the open load voltage $V_o(OL) = -A\ v_{ab}$, where A is the gain of the amplifier and v_{ab} is the voltage betwen terminals a and b. A load connected to the amplifier draws a current i_L from the amplifier. The output voltage is reduced by the voltage $i_L R_o$ across R_o, giving an output voltage v_o, under load

$$v_o = v_{o(OL)} - i_L R_o$$

Output resistances of operational amplifiers are on the order of a few tens or hundreds of ohms. In an ordinary application of the operational amplifier when peripheral feedback and input resistors are connected to the operational amplifier, the effective output resistance of the amplifer is only a few ohms, or a fraction of an ohm.

1.2.4 Common-Mode Gain (CMG) and Common-Mode Rejection Ratio (CMRR)

The operational amplifier ideally amplifies only the difference between the two signals applied to the inverting and noninverting input terminals. Ideally, if the two signals are equal in magnitude and in phase, they are not amplified at all, since their difference is zero. Such signals are called *common mode* signals. They can be caused, for example, by 60 Hz signals induced from neighboring equipment or circuits, or by the drift of a differential output stage of a transducer feeding the operational amplifier.

In practical operational amplifiers common mode signals are amplified to some degree since perfect matching of components inside the amplifier is impossible. This is called common mode gain (CMG). To minimize the effect of common mode signals, operational amplifiers are designed to have low common mode gain. The degree of interference seen at the output due to common mode input depends on the relative magnitudes of the differential and common mode input signals and on the *relative amplifications* of these signals. Accordingly, the *common mode rejection ratio* (CMRR) is defined as the absolute value of the ratio of the differential gain divided by the common mode gain:

$$CMRR = \left| \frac{A(w)}{CMG} \right| \qquad (1\text{-}1)$$

As an example, suppose that $|CMG| = 3$ and $|A(w)| = 100,000$ for low frequencies. Then

$$CMRR = \frac{100,000}{3} = 33,333$$

and in dB units

$$CMRR (dB) = 20 \log_{10} 33,333 = 90.46 \text{ dB}$$

Common-Mode Effect on the Output

The amplified common mode input signal appears at the output together with the amplified differential input signal, the former being amplified by the common mode gain, CMG, and the latter by the differential gain $-A(w)$. With reference to Figure 1-2, the output v_o, is given by the relationship

$$v_o = -A(w)(v_a - v_b) + (CMG) v_{cm} \qquad (1\text{-}2)$$

where v_a and v_b are respectively the voltages, measured with respect to ground, at terminals a and b, and v_{cm} is the common mode voltage appearing at terminals a and b, as measured with respect to ground. Solving Equation 1-1 for CMG and substituting the result in Equation 1-2 we obtain

$$v_o = -A(w)\left[(v_a - v_b) \pm \left| \frac{V_{cm}}{CMRR} \right| \right] \qquad (1\text{-}3)$$

which expresses the effect of the common mode signal on the output in terms of the common-mode rejection ratio CMRR which is specified by the manufacturer. Note that in Equation 1-3 the positive value of $|v_{cm}/CMRR|$ is preceded by a $\pm$ symbol as it is impossible to know *a priori* whether the common mode signal is amplified by a positive or negative gain. The sign can be different for two I.C.'s of the same type and model. It will be shown in Section 1.6 how common mode signals affect the output of practical amplifier configurations.

Maximum Common-Mode Input

There is an amplitude limit to common mode input voltages that may be applied to the inverting and non-inverting terminals. Beyond this limit the amplifier does not operate linearly. This limit is often given in the specification sheet for the amplifier. Maximum common mode input voltage is an important consideration in applications such as comparators where a reference signal is applied to one terminal and the output switches when the signal at the other input terminal just exceeds the reference signal, or in applications where only a single voltage supply is used and the quiescent voltages at the input terminals are above ground.

1.2.5 Voltage Supply Rejection Ratio (VSRR)

The characteristics of an operational amplifier are given for a certain set of test conditions including the voltage supplies such as ±15V. If the voltage supply drifts from its nominal value, the output voltage of the amplifier, in general, changes. The change in output voltage is expressed in terms of an equivalent differential input voltage which causes the same change in output voltage.

Thus, the effect of power supply voltage variation, i.e., the supply voltage rejection ratios, is given in terms of an equivalent input voltage per one volt change in the power supply voltage. The units are $\mu V/V$ or mV/V. A typical value is $20\mu V/V$. The supply voltage rejection ratio is also given in decibels. For example,

$$20 \log_{10} 20\mu V/V \quad = \quad 20 \log_{10} 20 \times 10^{-6} \quad = \quad -94 dB$$

Usually the absolute value, e.g., 94 dB, is given in the specification.

Depending on the degree of output stability required, the designer must choose a suitable operational amplifier and a suitable power supply. Obviously, the poorer the amplifier's voltage supply rejection ratio, the better the regulation and drift characteristics of the power supply must be.

Example. Consider an amplifier of the configuration shown in Figure 1-6a with $R_1 = 10$ KΩ and $R_2 = 50$ KΩ. The VSRR of the amplifier is specified at $200\mu V/V$. A voltage drop of 100 mV at the V+ terminal is caused by other components of the system. By how much will the output voltage change as a result of the change in power supply voltage?

Solution. The supply voltage drop of 100 mV or 0.1 V has the same effect on the output as the effect caused by 20μ or 20×10^{-6} V between the input terminals of the operational amplifier. The effect is calculated by multiplying this voltage by the quantity $(1 + R_2/R_1)$.

$$v_o = (1 + \frac{50 \times 10^3}{10 \times 10^3}) \quad 20 \times 10^{-6} = 6 \times 20 \times 10^{-6} = 120 \times 10^{-6} V$$

or $\quad 0.120$ mV

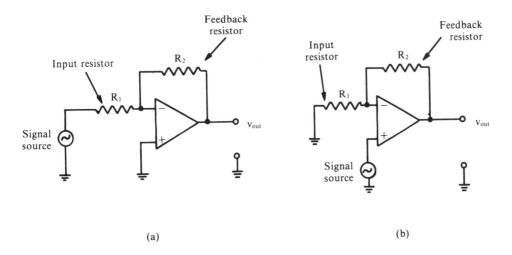

Figure 1-6. Basic operational amplifier applications.

(a) The inverting amplifier.
(b) The noninverting amplifier.

1.2.6 Input Offset Voltage

An ideal amplifier has zero output voltage for zero input voltage. Since real components cannot be perfectly matched, and since components in symmetrical positions may be at slightly different temperatures, the differential amplifier stages of an operational amplifier are, in general, not balanced perfectly. Consequently, there is a finite output voltage with zero input voltage. The *offset voltage* is defined as the differential voltage which must be applied to the *input terminals* to reduce the output voltage to zero. This voltage can be a few mV in magnitude. The polarity of the offset voltage is not known *a priori*. The *quiescent output voltage* resulting from the offset voltage depends on the closed loop amplifier design and is approximately equal to the offset voltage multiplied by the closed loop gain of the amplifier.

To balance the amplifier, a potentiometer is connected to a pair of balancing terminals as shown in Figure 1-5. The potentiometer is adjusted to obtain zero output voltage. A resistance value for the potentiometer is usually supplied by the manufacturer.

Since the offset voltage, like other characteristics of the amplifier, is temperature sensitive, the offset voltage can be balanced out only at a given temperature. It is therefore necessary to set the balancing potentiometer at the operating temperature, and check the output over the relevant temperature range.

1.2.7   Input Bias Current

D-C currents must be applied to both input terminals of an operational amplifier to bias the amplifier transistors properly. The circuitry which allows this bias current to flow is supplied externally to the operational amplifier and is the responsibility of the application engineer. The manufacturer specifies the magnitude of the bias current for the operational amplifier. In practical operational amplifiers the two input transistors are not matched perfectly, and in general the two bias currents I_{b1} and I_{b2} are different. The specified value I_b is the average of the two

$$I_b = \frac{I_{b1} + I_{b2}}{2}$$

Improper design of the bias circuit causes an undesirable output voltage which may exceed that caused by the input offset voltage.

To understand problems that can arise from improper bias circuitry and to learn how to alleviate such potential problems, we must pause for a moment and look at the two basic applications of the operational amplifier shown in Figure 1-6. The inverting amplifier has an output voltage which is 180° out of phase with the input signal voltage while the noninverting amplifier has an output voltage in phase with the input signal voltage. Each configuration offers certain unique features which are described in Sections 1.6.1 and 1.6.2. To consider the bias problem, we recall that to study d-c characteristics of an electrical circuit, a-c voltage sources are replaced with short circuits. Shorting the signal sources in Figure 1-6, both amplifier configurations take the form shown in Figure 1-7(a).

Because of the high gain of operational amplifiers and because of the negative feedback in the amplifier of Figure 1-7(a) there is a tendency of the potentials at the inverting input terminal a and the noninverting terminal b to be the same. If, say, the potential at a were higher than the potential at b, then there would be a negative output voltage. The output voltage is fed back through R_2 lowering the potential a. If the open loop gain of the operational amplifier were infinite, the difference between the potentials at a and b would be zero. Since terminal b is held at ground potential, we say that terminal a is at virtual ground. In first-order analyses, we can assume $v_a - v_b = 0$ where v_a and v_b are respectively the potentials (or voltages with respect to ground) at terminals a and b.

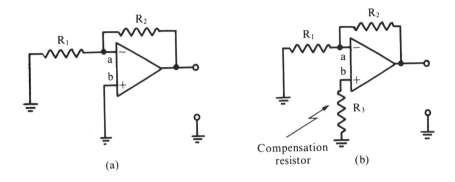

Figure 1-7. The amplifier of Figure 1-6 with short circuited voltage sources.

(a) Not compensated for bias imbalance.
(b) Compensated for bias imbalance.

We now study the bias problem. Since terminal a is at virtual ground potential, the bias current into terminal a must flow through R_2. (If bias current were to flow through R_1, there would be a voltage drop across R_1 causing the potential at a to be below ground, which contradicts the imposed condition that a is at ground potential.) To allow the bias current I_b to flow through R_2, the output voltage rises to a value of I_bR_2, equal to the voltage drop across R_2. The noninverting terminal b receives bias current from ground through the base emitter junction and the current flows to the negative supply terminal. We thus find that the circuit of Figure 1-7(a) has an output voltage I_bR_2 with no signal input. This is an undesirable condition as we want zero output with zero input. This undesirable condition is avoided by the addition of a compensation resistor R_3 as shown in Figure 1-7(b). The value of R_3 must be equal to the value of the parallel combination of R_1 and R_2.

$$R_3 \;=\; R_1 \| R_2 \;=\; \frac{R_1 R_2}{R_1 + R_2}$$

With R_3 in the circuit, the bias current into terminal b flows through R_3 producing a voltage at b of magnitude

$$v_b \;=\; -I_bR_3$$

Since as noted above $v_a - v_b = 0$, the voltage at terminal a is

$$v_a \;=\; -I_bR_3$$

This value in turn determines the currents through R_1 and R_2. Let I_b' be the current through R_1, and I_b'' the current through R_2. We have

$$I_b' = - \frac{v_a}{R_1} = \frac{I_b R_3}{R_1}$$

and, for zero output voltage,

$$I_b'' = \frac{-v_a}{R_2} = \frac{+I_b R_3}{R_2}$$

The total bias current into terminal a is

$$I_b' + I_b'' = I_b R_3 \left(\frac{1}{R_1} + \frac{1}{R_2} \right) = I_b R_3 \frac{R_1 + R_2}{R_1 R_2} = I_b R_3 \left(\frac{R_2 + R_1}{R_2 R_1} \right)$$

Since we chose $R_3 = \dfrac{R_1 R_2}{R_1 + R_2}$, therefore $R_3 \ \dfrac{R_1 + R_2}{R_1 R_2} = 1$ and

$$I_b' + I_b'' = I_b$$

Thus, we get the required bias current while maintaining zero output voltage as required. Thus, the choice of R_3.

   Example. Let I_b = 200 nA, R_1 = 2KΩ, and R_2 = 50KΩ. Then R_3 = $R_1 \| R_2$ = 1.9 KΩ, $v_a = v_b$ = –0.38mV, and v_o = 0 V. If R_3 were not used, we would have $v_a = v_b$ = 0 and $v_o = I_b R_2$ = 10 mV.

   The lowest permissible values of R_1, R_2, and R_3 are determined by input impedance and gain requirements as is made clear in sections in which the basic amplifiers are presented; the upper limit is determined by the input offset current discussed in the next section.

1.2.8   Input Offset Current

   In real operational amplifiers the bias currents I_{ba} and I_{bb} into terminal a and b, respectively, are not equal. The difference $|I_{ba} - I_{bb}|$ is the offset current. The offset current causes an output of the amplifier in the absence of an input. Consider Figure 1-7(b). Let R_2 = 500 KΩ and R_1 = 100 KΩ. We compute $R_3 = R_1 \| R_2$ giving R_3 = 83.3 KΩ. Let I_{bb} = 300 nA and I_{ba} = 400 nA. Then $v_b = -I_{bb} R_3$ = –2mV. But $v_a = v_b$, giving the current through R_1 as $I_b' = -v_b / R_1$ = 250 nA. The difference I_b'' between I_{ba} and I_b' flows through R_2. Thus $I_b'' = I_b - I_b'$ = 150 nA and $v_o = v_a + R_2 I_b''$ = 10mV. We see that the input offset current causes an undesirable output just as does the input offset voltage. Usually, the application engineer designs the circuit so that the effect of the offset current on the output is less than the effect of the offset voltage. The offset current puts an upper limit on resistor values that can be used to keep the output voltage within the prescribed specifications.

1.2.9 Gain-Bandwidth Product

The gain-bandwidth product of an operational amplifier is given by the product of the d-c gain of the amplifier and the 3 dB frequency. The three dB frequency is the frequency at which the gain is down by 3 dB from the d-c gain. Since the manufacturer does not know the closed-loop gain which will be employed by the user, the open-loop gain-bandwidth product is specified. This is a meaningful figure if it is known that the gain roll-off beyond the 3 dB point is -20 dB/decade (-6 dB/octave) as in Figure 1-3. Knowing this, it is possible to predict the gain-bandwidth product for any closed-loop gain. The intersection of the open-loop transfer curve and a straight line drawn parallel to the abscissa at a height equal to the closed-loop gain gives the 3 dB frequency. Multiplication of the closed-loop gain by this frequency gives the gain-bandwidth product. *Note*: For a roll-off of -20 dB/decade, the open-loop response curve intersects the 0dB (unity gain) line at a frequency equal to the gain-bandwidth product of the amplifier (1 MHz in the example of Figure 1-3).

When the open-loop response of an amplifier rolls off at a greater slope than 20 dB/decade, the open-loop gain-bandwidth product has no significance for closed-loop operation. In this case the manufacturer usually specifies the unity gain frequency which is somewhat more indicative of closed-loop capabilities, but in either case the usefulness of this information is limited and the complete frequency response curve is needed for closed-loop performance prediction. It is also important to keep in mind that the response curve is changed entirely if the amplifier is *compensated* for stability. Compensation is discussed in Section 1.4.

1.2.10 Slew Rate

The *slew rate* (SR) is defined as the highest possible rate of change for the amplifier output voltage.

$$SR = \left(\frac{dV_o}{dt}\right)_{max}$$

This specification is important for *large signal* operation. The slew rate of an amplifier is limited by the currents available to charge parasitic capacitances or deliberately introduced capacitors in the amplifier. At certain points in the circuit of an operational amplifier the response to a signal at the input terminals is dependent on a *current source*. This current charges the various capacitances building up the output voltage in response to the input voltage. The voltage v_0 across a capacitor is equal to the quotient of the charge q and the capacitance C,

$$v_c = q/C$$

For a constant charging current I_0, this equation becomes

$$v_c = \frac{I_0}{C} t$$

The rate of voltage change across the capacitor is obtained from this equation as

$$\frac{dv_c}{dt} = \frac{I_0}{C}$$

The slew rate (SR) is given by

$$SR = \frac{I_{max}}{C}$$

where I_{max} is the maximum current available in the circuit of the operational amplifier to charge the capacitors.

The slew rate imposes a frequency as well as an amplitude limitation. An amplifier may respond faithfully to an input signal of a certain frequency, but give a distorted output in response to an input signal of the *same* frequency when the output voltage called for is of a higher amplitude. A voltage of a given frequency and amplitude undergoes a higher rate of change than a voltage of the same frequency, but a lower amplitude, as is illustrated in Figure 1-8. The requirement for a larger output amplitude can be caused by a larger input signal and/or by a higher closed-loop gain of the amplifier. By similar reasoning it is also easy to see that of two signals of the same amplitude, the one of the higher frequency requires a larger rate of change.

The amplitude and frequency of a signal are related to the required slew rate (SR). Consider an output voltage of amplitude V_0 and frequency f (angular frequency $\omega=2\pi f$) given by

$$v = V_0 \sin\omega t$$

The rate of change is

$$\frac{dv}{dt} = \omega V_o \cos\omega t$$

and the maximum rate of change is given when $\cos\omega t = 1$,

$$\left(\frac{dv}{dt}\right)_{max} = 2\pi f V_o$$

Hence, an undistorted sine wave, of amplitude V_o and frequency f at the output, requires a slew rate, SR,

$$SR \geq 2\pi f V_o$$

For example, if an output of 7V peak at a frequency of 10KHz is required, an amplifier with a slew rate of at least 0.44 V/μs must be used.

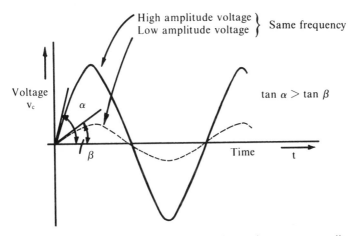

Figure 1-8. Rate of change of voltage as dependent upon amplitude.

1.2.11   Crossover Distortion

In the design of equipment using operational amplifiers it is usually assumed that the gain of the amplifier is linear over the operating range. This is not true for small signals near the zero crossover *if* the amplifier uses a *class B* output stage. Two requirements for an output stage are low output impedance and high power capabilities. An emitter follower output stage possesses these properties, but has the drawback of high power dissipation in the emitter resistor which causes heating of the amplifier chip and low efficiency of the amplifier. These shortcomings are removed if the output stage is operated *class B* using a complementary pair (npn and pnp) of transistors, but different shortcomings are introduced.

The zero crossover when one transistor turns on and the other transistor turns off is never perfect. The input-output characteristic of the amplifier is then as shown in Figure 1-9. The output is not only distorted, but also the open-loop gain (which is given by the slope of the curve) is considerably smaller for small signals than for larger signals. Since the zero crossing region is relatively narrow it is usually of no concern. But in applications where this degree of distortion or the small gain effect cannot be tolerated, the designer must avoid the use of an operational amplifier with a *class B* output stage.

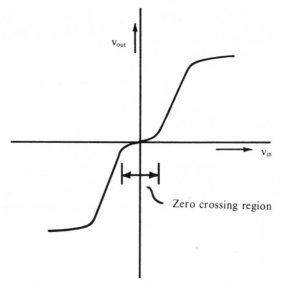

Figure 1-9. Input-output characteristic of *class B* amplifier.

1.2.12   Rated Output

The manufacturer ususally specifies maximum output voltage and maximum output current. The values represent limits above which the output is distorted, or the amplifier destroyed. If the voltage is exceeded, the amplifier enters saturation. This results in distortion and there is a lengthy recovery time. It should be noted that some amplifiers—the comparators— are used in saturated operation and are especially deisgned for relatively short recovery times.

Many operational amplifiers are built with current overload protection. The circuit operates in the normal mode for load currents below a specified point, and enters into the overload mode when the rated current is exceeded, resulting in distortion. The user must be aware that lack of overheating does not imply normal operation.

1.2.13   Power Dissipation

Two *power dissipation* values can be given: *device dissipation,* P_D, and *total dissipation,* P_T. The device dissipation is the power which can be safely consumed by the amplifier. The total dissipation refers to the amplifier plus load dissipation. Thus, permissible load power $P_L = P_T - P_D$. Any two values may be given by the manufacturer. The load power may be given in terms of maximum load voltage and maximum load current for resistive loads. The product of voltage and current gives power.

1.2.14   Input Overload Protection

Some amplifiers have an input overload protection. For differential input voltages below a specified value, the protection circuit is inactive. For larger differential input voltages, any excess current which would damage the input transistor is redirected to bypass the transistor resulting in distorted output. The designer must be careful to keep the signal values below levels that activate the protection circuit to maintain distortionless operation. If an application requires large differential input voltages, the designer must use an amplifier without input overload protection.

1.2.15   Supply Current Drain

Since normally more than one I.C. is supplied from one power supply, care must be taken that the total current drain does not exceed the power supply rating. Note that some amplifiers can be operated from a wide range of supply voltages and the current drain may depend on the voltage. In others, the current drain may be independent of the supply voltage.

1.2.16   Transient Response

When a pulse is applied to the input of the amplifier, the output responds as shown in Figure 1-10. The manufacturer may specify the percentage *overshoot, rise time* (10% to 90% of final output) and *settling time,* as well as slew rate discussed in Section 1.2.10. All these specifications are defined in the figure.

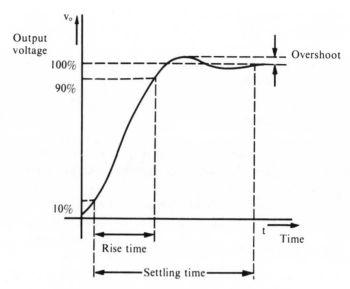

Figure 1-10. Transient response of amplifier.

1.2.17   Input Capacitance

The parasitic input capacitance is usually in the range of one to ten picofarads and is specified by the manufacturer.

1.2.18   Amplifier Noise

Amplifier noise is an important consideration in low level signal application. The complete design must consider noise orginating from the amplifier as well as other circuit components. The subject of noise and how to optimize designs for minimum noise is discussed in detail in Chapter 3.

1.2.19   Temperature, Frequency, Supply Voltage, Source Resistance and Load Resistance Dependencies of Amplifier Characteristics

The characteristics of an operational amplifier are specified under certain *operating* or *test* conditions. These conditions may include load resistance, temperature and other parameters. So that the reader can study orders of magnitudes, units, and parameter dependencies of the various characteristics, a complete set of specifications of the bipolar 741[1] and the FET input LH0042 amplifiers are included at the end of this chapter.

---

[1]The 741 characteristics are given by the 747 data sheets. The 747 is a package containing two 741's. The characteristics are identical.

1.3   DUAL and SINGLE POWER SUPPLY AMPLIFIERS

Most operational amplifiers are designed for use with dual voltage power supplies (±15 V or ±12 V are typical). This permits input and output signals to swing above and below ground voltage without the need for coupling capacitors. However, a dual voltage power supply can be more expensive than a comparable single voltage power supply of twice the voltage. Amplifiers usually supplied by a dual voltage power supply can be operated from a single voltage power supply, in which case an external voltage divider must be added to set the input half way betwen the supply voltage extremes. As an example, the CA3015 (RCA) amplifier normally operates from a ±12 V supply. It can also be operated from a 24 V supply. A diagram showing the power supply connections is shown in Figure 1-11 (a). *Note:* The input and output are capacitively coupled to the amplifier.

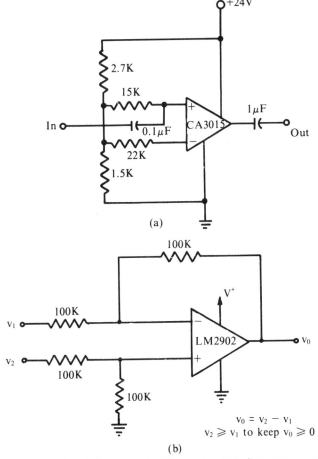

(a)

$$v_0 = v_2 - v_1$$
$$v_2 \geqslant v_1 \text{ to keep } v_0 \geqslant 0$$

(b)

Figure 1-11. Amplifier operated from single voltage power supply.

(a) CA 3015   (b) LM 2902

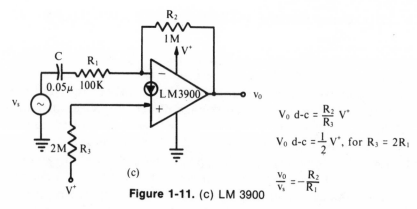

$$V_0 \text{ d-c} = \frac{R_2}{R_3} V^+$$

$$V_0 \text{ d-c} = \frac{1}{2} V^+, \text{ for } R_3 = 2R_1$$

(c)

Figure 1-11. (c) LM 3900

$$\frac{V_0}{V_s} = -\frac{R_2}{R_1}$$

With the increased interest in applications of I.C.'s in the automotive industry, operational amplifiers are available explicitly designed for use with a single voltage power supply. Some can also be used with a dual voltage power supply. For example, the LM2902 quad operational amplifier (National Semiconductor) contains four operational amplifiers in one package. These amplifiers can be operated from a wide range of power supplies from 3 V to 26 V, or from ± 1.5 V to ± 13 V. Thus, the amplifier can be operated from a 5 V power supply in a digital system, eliminating the need for a separate dual voltage power supply. This amplifier can be used to sense signals near ground. An example showing the application of the device in a difference amplifier is shown in Figure 1-11(b).

A popular single polarity power supply amplifier is the LM3900 quad operational amplifier (National Semiconductor). This amplifier is different from the conventional amplifiers in that the input stage is not a voltage differential amplifier, but a current differential amplifier. The inverting input stage is a common emitter stage. The non-inverting input stage is provided by a circuit known as a *current mirror* circuit. This results in an input stage in which currents are compared or differenced. It is called a *Norton* differential amplifier. There is no limit to the common mode input voltage range. This is useful in high voltage comparator applications. By making use of input resistors, to convert input voltages to input currents, all of the standard operational amplifier applications can be realized. An application example is shown in Figure 1-11(c), where we show an inverting amplifier. Note that the d-c level at the output is determined by a resistance ratio. In particular, the d-c level can be set half way between the power supply voltage and ground by choosing $R_3 = 2R_2$. An important practical advantage of this amplifier is that no voltage dividing network is needed to establish the bias level of the input stage, which would reduce the input impedance. The non-inverting input is simply connected to the power supply through a resistor. Note that the Norton amplifier is identified by two arrows, one between the inverting and non-inverting terminals, and one into the non-inverting terminal.

1.4   PHASE COMPENSATION

Amplifier instabilities and oscillations occur if a voltage is fed back from the output to the input with a total loop phase shift of 360° and unity gain. Oscillations can be initiated by noise anywhere in the circuit. The right conditions for oscillations usually occur in the 10 kHz to 30 MHz region. To prevent such oscillations from occurring the amplifier is *compensated* by means of RC networks. These networks modify the open loop frequency response of the amplifier which in turn results in stable closed-loop operation. The simplest compensation method consists of reducing the overall gain of the amplifier with resistive networks; this, of course, defeats the main desirable characteristic of operational amplifiers, high open-loop gain. A practical method consists of reducing the gain in certain frequency ranges. This is always associated with a change in the phase shift-frequency response of the amplifier. In fact, stability criteria have evolved about phase shifts and *phase margins* . Since, however, the amplitude-frequency and phase shift-frequency characteristics are intrinsically related (both being dependent on the same resistive and capacitive components), the necessary information for our purposes, can be obtained directly from the amplitude-frequency characteristic.

Stability is assured if the *closed-loop* gain transfer function meets the

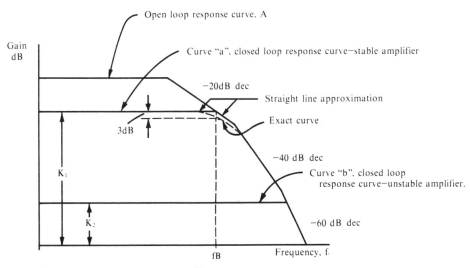

For the inverting amplifier K_1, $K_2 = \frac{-R_2}{R_1}$ (See Fig. 1-6a)

For the noninverting amplifier K_1, $K_2 = 1 + \frac{R_2}{R_1}$ (See Fig. 1-6b)

Figure 1-12. Response curves of stable and unstable amplifier.

open-loop gain transfer function where the open-loop tansfer function has a slope of –20 dB/decade. Figure 1-12 shows the open-loop transfer curve of an amplifier and closed-loop tansfer curves for two different feedback ratios. Curve "a" corresponds to a stable amplifier, while curve "b" corresponds to an unstable amplifier. It is seen that the stable amplifier incorporates less feedback (i.e. has higher closed-loop gain) than the unstable amplifier. Given the open-loop response curve of an amplifier, the designer can determine the closed-loop gain range for which the amplifier is stable. The bandwidth of the stable amplifier is given by f_B, the frequency at the 3 dB attenuation point. The lower the closed-loop gain, the wider the bandwidth.

1.4.1   Internally Compensated Amplifiers

The obvious advantage of amplifiers that are internally compensated by the manufacturer is the convenience given to the application engineer not to have to compensate the amplifier. The disadvantage is that these amplifiers have very low cut-off frequencies. Since it is the intention to have an internally compensated amplifier stable under all probable applications, the open loop is modified to have a –20 dB/dec slope throughout the complete frequency range, starting from the break frequency of about 5Hz as shown by the internally compensated line in Figure 1-13. Lines A, B, and C in the figure show, for example, that the amplifier is stable for gains of 20,

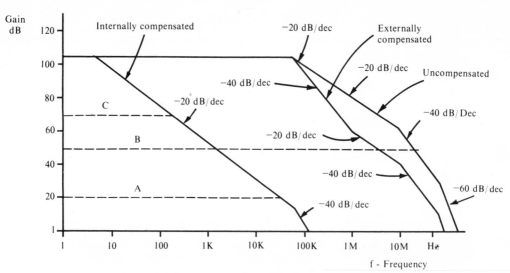

Figure 1-13. Amplifier compensation — uncompensated, externally compensated, and internally compensated gain-frequency response curve.

50 and 70 dB since for each of these gains the closed-loop curve meets the open-loop-gain curve where the latter has a –20dB/dec slope. Note that the 3 dB frequencies are at 60 kHz, 2 kHz, and 150 Hz respectively.

1.4.2 Uncompensated Amplifiers

Uncompensated amplifiers are more versatile since they can be compensated to give closed-loop gains with a wider frequency response than an internally compensated amplifier designed for the same closed-loop gain. The compensation responsibility lies however with the application engineer.

Consider the uncompensated gain frequency curve shown in Figure 1-13. If the uncompensated amplifier is used for a closed-loop gain of 50 dB, the closed-loop gain-frequency curve meets the open-loop curve where the latter has a slope of –40 dB/dec and the amplifier is likely to be unstable. If the open-loop curve is reshaped as shown by the "externally compensated" curve, the amplifier is stable with a closed-loop gain of 50 dB because the closed-loop curve meets the compensated open-loop curve at a −20 dB/dec

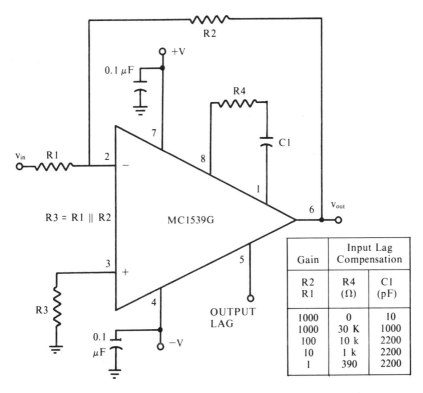

Gain	Input Lag Compensation		
$\frac{R2}{R1}$	R4 (Ω)	C1 (pF)	
1000	0	10	
1000	30 K	1000	
100	10 k	2200	
10	1 k	2200	
1	390	2200	

Figure 1-14. External compensation of amplifier. (a) Circuit Diagram. (Reprinted with permission from Motorola Semiconductor Products, Inc.)

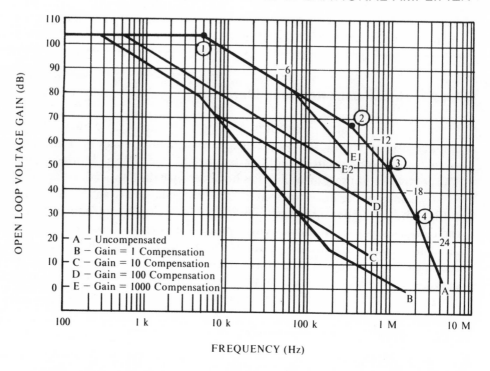

Figure 1-14. (b). Response Curves. (Reprinted with permission from Motorola Semiconductor Products, Inc.)

slope of the latter. The amplifier has an 8 MHz 3 dB point as compared to the 2 Khz 3 dB point of the internally compensated amplifier. Note that the externally compensated amplifier is likely to be unstable for a closed-loop gain of 20 dB as well as 70 dB. This example illustrates very clearly how the externally compensated amplifier can be custom tailored for a particular application, while the internally compensated amplifier cannot.

The uncompensated curve can be reshaped to that of the externally compensated curve by connecting an appropriate network to the "compensation" terminals of the amplifier. The required procedure for compensation for practical amplifiers is supplied by the manufacturer. For example, Figure 1-14 gives compensation instructions in the form of diagrams, tables, and response curves for the MC1359 operational amplifier (Motorola). The information gives the required values of the compensating network components R_4 and C_1 for particular values of desired closed loop gain. The two $0.1 \mu F$ capacitors shown in the figure are power supply decoupling capacitors removing variations in the power supply caused by other devices in the system. It is good practice to use such decoupling capacitors for all I.C.'s in a system.

1.5   TRANSCONDUCTANCE OPERATIONAL AMPLIFIER

Whereas the basic characteristic of the voltage operational amplifier, discussed so far, is the forward voltage gain, the basic characteristic of the transconductance amplifier is the *transconductance* g_m, that is, the ratio of the amplifier output current i_{out} to the input voltage v_{in}: $g_m = i_{out}/v_{in}$. The output of the amplifier is a current proportional to the input voltage. This is accomplished by providing a very large output impedance in contrast to a very low output impedance in voltage amplifiers. The transconductance amplifier is a high impedance circuit; it is meant for low currents (microamps) and low dissipation of a few mW, or a fraction of a mW and as little as a few nW. Of particular interest is the provision which allows the designer to adjust the transconductance of the amplifier. This is accomplished by providing the appropriate amplifier bias current I_{ABC}. In practice this is accomplished simply by the selection of the value of a bias resistor R_{ABC}. An inverting amplifier circuit which uses a transconductance amplifier is shown in Figure 1-15a. The current through R_L is proportional

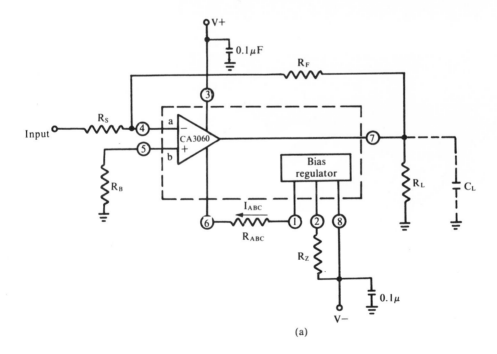

Figure 1-15. Inverting transconductance amplifier

(a) Schematic

(Reprinted by permission of RCA Solid State Division)

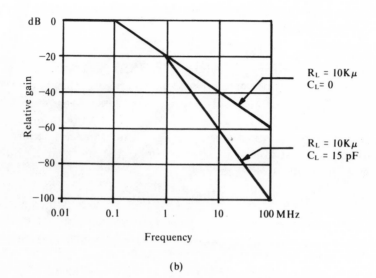

(b)

Figure 1-15 (continued)

(b) Effect of capacitive loading on frequency response

to the voltage applied to the input terminal. The design procedure for the amplifier is given in Example 2 of Section 1.8.

The characteristics of the I.C., including *input offset voltage, input bias current, peak output current, peak output voltage, forward transconductance, and output resistance* are determined by the amplifier bias current. A set of curves which give the characteristics of the 3060 (RCA) is given at the end of this chapter.

1.6   BASIC VOLTAGE AMPLIFIERS

In this section we introduce fundamental linear applications of the operational amplifier. Additional applications are given in Chapter 5.

1.6.1   The Inverting Amplifier

The inverting amplifier configuration is shown in Figure 1-16. Resistors R_1 and R_2 determine the voltage gain of the amplifier; R_1 determines also the input resistance, and R_3, set equal to the parallel combination of R_1 and R_2, eliminates imbalance due to bias current, but not due to unequal bias currents as was discussed in Section 1.2.6.

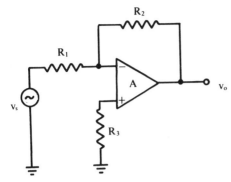

Approximate equations for practical design:

$$\text{Gain} = -\frac{R_2}{R_1}$$

Input resistance $= R_1$

Output resistance $= R_o \dfrac{R_2}{AR_1}$

Figure 1-16. The inverting amplifier.

Voltage Gain

The voltage gain of the amplifier is:

$$\frac{v_o}{v_s} = \frac{A}{1 + \dfrac{R_1}{R_2}(1+A)}$$

where v_o and v_s are respectively the output and input voltages, A is the open loop operational amplifier gain, and R_1 and R_2 are the external input and feedback resistors, respectively.

Note that the closed-loop voltage gain (v_o/v_s) is a function of frequency since the open-loop gain A is a function of frequency. (See Figures 1-3 and 1-12.) At low frequencies where $(R_1/R_2)(1+A) \gg 1$ and $A \gg 1$ the closed-loop gain equation becomes for all practical purposes

$$\frac{v_o}{v_s} = -\frac{R_2}{R_1}$$

This is usually the frequency range in which the amplifier is used and therefore this last equation is the common design equation. Practical constraints limit permissible values of R_1 and R_2 making an arbitrary large gain impossible. Required input impedance imposes a lower limit on R_1 and offset bias current limits upper values of both R_1 and R_2. These constraints are taken into consideration in design example 1, Section 1.8.

At high frequencies for which $(R_1/R_2)(1+A) \ll 1$, the closed-loop voltage gain becomes $(v_o/v_s) = -A$. That is, at high frequencies the closed-loop voltage gain becomes equal to the open-loop voltage gain. This is illustrated in Figure 1-12. The magnitude of the gain changes from R_2/R_1 to A at a frequency f_B at which the corresponding "straight line approximation" lines intersect. In reality, the transition from one gain to the other is gradual and at f_B the gain is down approximately by 3 dB from the low frequency gain. (If the open-loop gain curve is composed of two linear sections of 0 dB/dec and -20 dB/dec as exemplified by the response curve in Figure 1-3, then at the intersection of the "straight line approximation" of the closed-loop and open-loop lines the gain is down *exactly* by 3 dB.)

Input Resistance

In practical designs the input resistance of the amplifier is equal to the resistance of the externally connected resistor R_1.

$$R_{in} = R_1$$

For this to be true the open-loop gain of the operational amplifier must be tens of thousands or larger; the open-loop output resistance of the operational amplifier must be about a hundred ohms or less; and the open-loop input resistance of the amplifier (i.e., the resistance between the two input terminals), must be at least two orders of magnitude larger than R_1. All these conditions are commonplace with I.C. operational amplifiers.

Output Resistance

For the values of open-loop gain and input resistance of practical operational amplifiers, the output resistance of the inverting amplifier is

$$R_{out} = \frac{R_o \left(1 + \dfrac{R_2}{R_1}\right)}{A}$$

where R_o is the open-loop output resistance of the operational amplifier. The value of R_o is typically on the order of a hundred ohms. Typically A is in the hundreds of thousands or millions, and R_2/R_1 is on the order of 10 or 100. We see that R_{out} is a small fraction of R_o. The output resistance can be approximated in terms of the *loop gain* $A_{Loop} = A(R_1/R_2)$ as

$$R_{out} = \frac{R_o}{A_{Loop}}$$

For $R_o = 200\ \Omega$, $A = 100,000$, and $R_2/R_1 = 10$, this give $R_{out} = 0.02\ \Omega$. Caution must however be exercised since R_{out} can increase significantly if the loop gain for any reason (e.g. high frequency) goes down. For example, if $R_o = 200\ \Omega$, $A = 10,000$, and $R_2/R_1 = 200$, then $R_{out} = 4\ \Omega$.

Common Mode Effect

The output of the amplifier including the effect of the common mode input is

$$V_o = -\frac{R_2}{R_1}\left[v_s + (1 + \frac{R_1}{R_2})\frac{v_{cm}}{CMRR}\right]$$

The expression $1 + R_1/R_2$ is close to unity showing that the common signal appears attenuated by a factor of CMRR before it is amplified by the gain of the amplifier. The output voltage equation is correct for large open-loop gain A. At high frequencies for which A is small, the gain of the closed-loop amplifier approaches $-A$ (rather than $-R_2/R_1$ given in the equation), and the relative effect of v_{cm} versus v_s increases since CMRR decreases with decreasing A.

The common mode rejection property reduces the effect of spurious or undesired signals reaching both input terminals of the operational amplifier. It does not help to reject undesired signals originating in the signal source or reaching the terminal to which the signal source is connected, as such signals do not possess the *common mode* aspect, but appear as input signals.

1.6.2   The Noninverting Amplifier

The noninverting amplifier configuration is shown in Figure 1-17.

Voltage Gain

The voltage gain of the amplifier is

$$\frac{v_o}{v_s} = \frac{A}{1 + \dfrac{R_1 A}{R_1 + R_2}}$$

where all the symbols are defined in the figure. The gain is a function of frequency since A is a function of frequency. (See Figures 1-3 and 1-12). At low frequencies for which $R_1 A/(R_1+R_2) \gg 1$, the gain equation becomes

$$\frac{v_o}{v_s} = 1 + \frac{R_2}{R_1}$$

The gain in this range cannot be less than unity. The amplifier is ordinarily used in a frequency range where this equation is true, making the equation the standard design equation. Note, however, that at high frequencies where

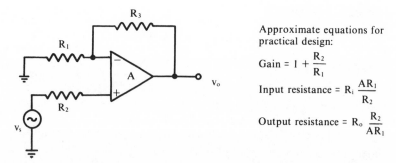

Approximate equations for practical design:

$$\text{Gain} = 1 + \frac{R_2}{R_1}$$

$$\text{Input resistance} = R_i \frac{AR_1}{R_2}$$

$$\text{Output resistance} = R_o \frac{R_2}{AR_1}$$

Figure 1-17. The noninverting amplifier.

$R_1 A/(R_1+R_2) \ll 1$, the gain of the amplifier becomes A equal to the open-loop gain of the operational amplifier. The transition from one gain to the other occurs at frequency f_B at which $A = 1+R_2/R_1$ as shown in Figure 1-12. The discussion concerning the transition given for the inverting amplifier can be adapted to the noninverting amplifier.

Input Resistance

For the practical conditions of high gain A, high open-loop input resistance and low output resistance of the open-loop operational amplifier, the input resistance of the noninverting amplifier is

$$R_{in} = \frac{R_i A}{1 + \dfrac{R_2}{R_1}}$$

This can be written as

$$R_{in} = R_i A_{Loop}$$

where A_{Loop} is defined as the loop gain equal to AR_1/R_1+R_2. In the frequency region where the amplifier is ordinarily used, the loop gain is very large and the input impedance R_{in} of the noninverting amplifier is several orders of magnitude larger than the open-loop input resistance R_i of the amplifier. Extraordinary high input resistances can be implemented. For example, if $R_i = 10^{12}\Omega$ and $A = 200,000$ (typical for FET operational amplifiers) and the closed-loop gain of the amplifier is 100, then the loop gain of the amplifier is $200,000/100 = 2,000$ and the input resistance is 2×10^{15} Ω. At high frequencies the open-loop gain decreases causing the input impedance to decrease, as can be computed from the expression for R_{in} given above.

Output Resistance

For the practical conditions of very large A and R_i the output resistance is

$$R_{out} = \frac{R_o \left(1 + \frac{R_2}{R_1}\right)}{A}$$

This is the same expression as for the output resistance of the inverting amplifier. Everything that was said there is also true for the noninverting amplifier.

Common Mode Effect

The output of the amplifier, including the effect of the common mode input is

$$v_o = \left(1 + \frac{R_2}{R_1}\right)\left(v_s + \frac{v_{cm}}{CMRR}\right)$$

for high open-loop gain A. For low values of A, (e.g. at high frequencies) the gain of the noninverting amplifier is A, not $1+R_2/R_1$ as given in the equation. The effect of v_{cm} becomes more pronounced as CMRR decreases with decreasing A. As was also true for the inverting amplifier, the common mode rejection property reduces the effect of spurious or undesirable signals reaching both input terminals of the operational amplifier, but not spurious or undesirable signals orginating in the signal source or reaching the terminal to which the signal source is connected as such signals do not possess the *common mode* aspect, but appear as input signals.

1.6.3   Choice between Inverting and Noninverting Configurations

If gain less than unity is needed, the inverting amplifier must be used since the gain of the noninverting amplifier is either equal to, or greater than, unity. If a very high input resistance is needed, the noninverting amplifier must be used. Its input resistance is equal to the product of the open-loop input resistance R_i of the operational amplifier and the loop gain of the amplifier, both of which have high values. The input resistance of the inverting amplifier is equal to the resistance connected between the signal source and the inverting terminal. If this resistance is chosen high to have a high input resistance, and the feedback resistance for a gain greater than unity, is even higher, the resistance values reach magnitudes too high for several reasons: they may introduce intolerable noise and cause unbalance due to unbalanced bias currents.

For multiple input signals the inverting amplifier is normally more useful. For two input signals, as shown in Figure 1-18a the output is

$$V_o = -R_3 \left(\frac{V_{s1}}{R_1} + \frac{V_{s2}}{R_2} \right)$$

If $R_2 = R_1$ then

$$V_o = -\frac{R_3}{R_1} (v_{s1} + v_{s2})$$

Each input contributes its share to the output, independent of the other input circuit. The same concept holds true for three or more input signals. The output of the noninverting amplifier (Figure 1-18b) is

$$V_o = \left(1 + \frac{R_2}{R_1}\right) \left(\frac{R_3 R_4}{R_3 + R_4}\right) \left(\frac{V_{s1}}{R_3} + \frac{V_{s2}}{R_4}\right)$$

The contribution to the output of an input signal is determined not only by its own circuit, but also by the circuit associated with the other input signal. For example, the output associated with V_{s1} depends not only on R_1, R_2, and R_3, but also on R_4, which is associated with the circuit of V_{s2}. The resistor R_4 may represent in addition to an external resistor also the output resistance of the signal source V_{s2}. Thus, the factor by which a signal is amplified depends on other signal sources connected to the amplifier.

If $R_3 = R_4$, then

$$V_o = \left(1 + \frac{R_2}{R_1}\right) \frac{1}{2} (v_{s1} + v_{s2})$$

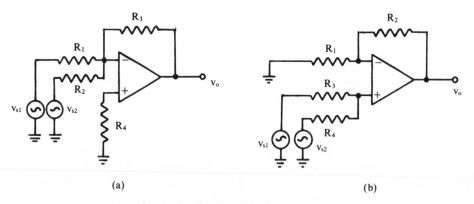

(a)   (b)

Figure 1-18. Multiple input amplifiers.
(a) Inverting amplifier
(b) Noninverting amplifier

The gain of the amplifier is only one-half the gain for a single input. If there are three input signals, the gain is reduced by 3, and if there are n input signal sources, the gain is reduced by a factor of n. This gain reduction does not exist in the inverting amplifier. Further, the high input resistance aspect of the noninverting amplifier is lost since the input resistance $R_{in}A_{Loop}$ is now paralleled with the resistors associated with the other input signal sources.

1.6.4 The Differential Amplifier

The schematic diagram of a differential amplifier is shown in Figure 1-19a. The amplifier accepts signals simultaneously at both the inverting and noninverting input terminals. A unique property of the differential amplifier is that it rejects common mode signals originating at the signal source. It thus rejects 60 cycle-induced voltages picked up by electrodes pressed against the skin in biological or medical measurements; it rejects signals resulting from drifts in a previous amplification or transducer stage; and it rejects unwanted signals originating in so-called ground loops. The output signal is the amplified difference between the two input signals,

$$v_o = - \frac{R_2}{R_1} (v_{s1} - v_{s2}) = - \frac{R_2}{R_1} v_s$$

where v_s is the voltage applied between the two input terminals and we set $v_{s1} = v_s/2$ and $v_{s2} = -v_s/2$.

In Figure 1-19a equal sets of resistors are chosen for both input

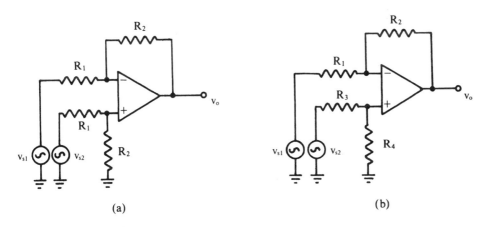

(a)   (b)

Figure 1-19. Differential amplifier.
(a) Signals weighted equally
(b) Signals weighted unequally.

circuits. In general, the resistors can be different (Figure 1-19b) resulting in an output of the weighted input signals.

$$v_o = \frac{R_4(R_1+R_2)}{R_1(R_3+R_4)} v_{s2} - \frac{R_2(R_3+R_4)}{R_1(R_3+R_4)} v_{s1}$$

This equation reduces to the previous one for the special case
$R_3 = R_1$ and $R_4 = R_2$.

Common Mode Effect

The differential amplifier rejects common mode signals appearing at the operational amplifier input terminals as well as those originating or appearing at the signal source. The output in the presence of a common mode voltage at the input terminals of the operational amplifier is

$$v_o = -\left[\frac{R_2}{R_1}(v_s) + (1 + \frac{R_2}{R_1})\frac{v_{cm}}{CMRR}\right]$$

For a common mode voltage at the source the output voltage is

$$v_o = -\frac{R_2}{R_1}(v_s + \frac{v_{cm}}{CMRR})$$

The relative contribution of the common signal v_{cm} to the output is small because it is divided by CMRR. As the frequency increases, the CMRR deteriorates (decreases) and the relative contribution of v_{cm} increases.

1.7   THREE-RESISTOR FEEDBACK NETWORK

An inverting amplifier with a three-resistor feedback network is shown in Figure 1-20a. The equivalent feedback resistance R_2 of this amplifier is

$$R_2 = \frac{R_a R_b + R_a R_c + R_b R_c}{R_c}$$

The voltage gain of the amplifier in terms of the equivalent resistance is

$$\frac{v_o}{v_s} = -\frac{R_2}{R_1}$$

The resistor R_3, as usual, is used to balance bias current effects and is equal to the parallel combination of R_1 and R_2. The advantage of using such an arrangement is illustrated by the following example.

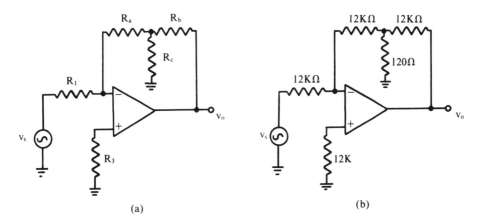

Figure 1-20. Inverting amplifier with three-resistor feedback circuit.

(a) General schematic diagram
(b) Example.

Example. Determine the voltage gain of the amplifier shown in Figure 1-20b.

Solution. The effective feedback resistance is

$$R_2 = \frac{12 \times 10^3 \times 12 \times 10^3 + 12 \times 10^3 \times 120 + 12 \times 10^3 \times 120}{120}$$

$$= 1{,}224 \times 10^6 \text{ ohms}$$

The voltage gain is

$$\frac{V_o}{V_s} = -\frac{1.224 \times 10^6}{12 \times 10^3} = -102$$

Had we used the amplifier configuration of Figure 1-16 with $R_1 = 12 \text{ K}\Omega$ (to maintain the same input impedance), then to have a gain of 102, the feedback resistor R_2 in Figure 1-16 would have to be $12 \text{ K} \times 102 = 1.224 \text{ M}\Omega$. The advantage of the circuit of Figure 1-20 is evident in that the highest resistor value of $12 \text{ K}\Omega$ is used in the circuit as compared to $1.224 \text{ M}\Omega$ to obtain the same gain and the same input impedance. The use of relatively low resistance values reduces the noise associated with resistors and also suppresses the effect of stray wiring capacitances.

1.8   DESIGN EXAMPLES

Example 1.   Voltage Operational Amplifier

It is required to design an amplifier to change the voltage level of a signal of maximum rms value of 300mV supplied by a source with 5 Ω output impedance. The maximum load current that the signal source can supply is 500 μA. The output of the amplifier must have a maximum rms value of about 6 V feeding into a 5 KΩ load. The maximum frequency of the signal is 18 kHz. The amplifier must operate over a temperature range of 0-50° C with a distortion not to exceed 1% of maximum output.

The statement that the amplifier output must be "about" 6 V means that the actual gain need not have a precise, specified value (perhaps an adjustable gain exists elsewhere in the system) which implies that the resistors which determine the gain need not be precision resistors; 5% resistors are adequate. We shall try to design the amplifier around the basic configuration of Figure 1-16. The resistance ratio R_2/R_1 which determines the gain of the amplifier must be 6V/300mV = 20. In decibels the gain is 20 log 20 = 26 dB. We consider the popular amplifier 741 (see data sheets at end of chapter).[1] An examination of the open-loop transfer characteristic shows for a closed-loop gain of twenty, a 3 dB frequency of about 50 KHz, and a peak-to-peak output swing at 18k Hz of 16V, while our need is for $2 \times \sqrt{2} \times 6 =$ 16.97 V peak-to-peak output voltage. Furthermore, a check of the required slew rate shows that we need SR$\geqslant 2\pi$ f V$_o = 2\pi$ 18 $\times$ 10^3 $\times$ 6 $\times$ 10^{-6} = 0.68 V/μs. The typical slew rate of the 741 is 0.5/μs. Thus, the use of this operational amplifier is ruled out. In terms of operational amplifiers the frequency requirement is severe, and an uncompensated amplifier may be the right choice. Examining the 748 (see data sheets at end of chapter), shows external compensation affected by means of a capacitor connected between two terminals. For a 3 pF capacitor the 3 dB frequency for a 26 dB closed-loop gain is about 500 kHz, and the output peak-to-peak voltage at 18 kHz is 29V. The slew rate of the amplifier is 0.8 V/μs. These values are for a power supply of $\pm$ 15 V and a temperature of 25° C, as read off the curves. The minimum gain over the range –55° C to 125° C is 93 dB. A shift of the gain frequency-response curve from the typical 100 dB (d-c) gain to 93 dB shows that the amplifier still meets our requirements, and there is ample margin since the required temperature range is limited to 0°–50° C. The output current swing for a peak-to-peak output voltage of 17 V is about 52 mA peak-to-peak at 25° C and 32 mA peak-to-peak at 125° C. Our need is for only 2 $\times \sqrt{2} \times$ (6/5000) = 0.0034A or 3.4 mA peak-to-peak. We use, therefore, the 748.

One percent of 6 V is 60 mV. Therefore, to ensure that the distortion not exceed 1% of maximum output, unwanted voltages at the input must be limited to 60 mV/20 = 3 mV. The *maximum* input offset voltage over the

----

[1]Identical to 747 characteristics.

range of $-55°$ C to $125°$ C for the 748, and $0°$ to $70°$ C for the 748 C is specified as 6.0 mV and the average coefficient of input offset voltage is 3.0 $\mu V/°C$. Thus, if we balance the amplifier to zero output at $25°$ C, then at the extremes of $0°$ C and $50°$ C the offset voltage will be $3\,\mu V°C \times 25\,C° = 75\,\mu V$, which is considerably below the 3 mV limit. This provides a high degree of confidence that the design will be valid although the 3.0 $\mu V/°C$ figure given in the data sheet is an *average* value. (A curve is not provided by the particular manufacturer.)

The minimum value of the resistor R_1 is imposed by the signal source specification of 300 mV and 500 μA: $R_{1\ min} = 300 \times 10^{-3}\ V/500 \times 10^{-6}\ A = 600\ \Omega$. The maximum value is determined by the input offset current. The error introduced by the bias current is compensated for by adding a resistor R_3 between the noninverting terminal and ground. However, the two bias currents are not exactly the same, and for the temperature range $0°$ C to $70°$ C the difference in current, i.e., the offset current, is specified as at most 300 nA. We can express this in terms of a voltage by noting that the bias current flows through R_1 and R_2 in parallel:

$$\frac{v_a}{R_1} + \frac{v_a}{R_2} = i_{offset}$$

$$i_{offset}\,\frac{R_1 R_2}{R_1 + R_2} = v_a$$

but $R_2 = 20\ R_1$, giving

$$i_{offset} \times 0.95\ R_1 = v_a$$

As a guide, we want to ensure that $v_a \leqslant 0.3$ mV, which is 10% of maximum total unwanted voltage permissible to ensure not more than 1% distortion of maximum output. Thus, for the upper limit,

$$300 \times 10^{-9} \times 0.95\ R_1 = 0.3 \times 10^{-3}$$

from which

$$R_1 \leqslant 1053\ ohms$$

Thus, we choose $R_1 = 1000\ \Omega$ and $R_2 = 20\ K\Omega$. These are standard 5% resistors in the required range giving a gain of 20. To balance the bias current, we connect a resistor $R_3 = 910\ \Omega$ (a standard 5% resistor approximately equal to the parallel combination of R_1 and R_2) to the noninverting terminal. The voltages at the input terminals resulting from the bias currents are given by $I_{bias} \times R_3 \times 0.3$ mV. This is small in comparison with the specified* common mode voltage swing of ± 12 V.

---

*Obtained from a manufacturer's data sheet.

The minimum common mode rejection ratio of the operational amplifier is specified as 70 dB or 3,162. From the common mode effect equation in Section 1.6.2 we find that a common mode signal of 0.3 mV at the input terminals of the operational amplifier contributes to the output.

$$(1 + \frac{R_2}{R_1})\frac{V_{cm}}{CMRR} = \frac{21}{3162} \, 0.3 \times 10^{-3} = 2 \times 10^{-6} \text{ V}$$

This is $(2 \times 10^{-6}/6) \times 100 = 3 \times 10^{-5}\%$ of maximum output.

The resistance values of the resistors used in the circuit are small enough for resistor noise not to pose a problem.

The complete circuit of the amplifier is shown in Figure 1-21. The voltage offset balancing circuit, consisting of P, R_4 and R_5 (all carbon type components), is taken from the data sheet. The 3 pF, ceramic capacitor is used for open loop gain-frequency compensation (or phase compensation) for stability as given in the data sheet. The 0.1 μF capacitors are tantalum electrolytic decoupling capacitors to filter any disturbances from other parts of the circuit which could be transmitted through the power supplies.

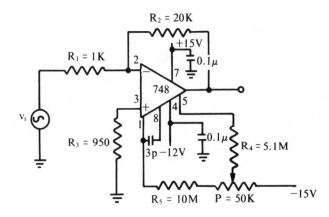

Figure 1-21. Design Example.

Example 2.   Transconductance Operational Amplifier

It is required to design an amplifier with a closed-loop voltage gain A_{CL} of 6, and an input resistance R_s of 25 KΩ to drive a load resistance R_L of 10 KΩ. The maximum input signal voltage is 100 mV peak-to-peak. The power supply voltages are ± 6 V and the temperature is 25°C. The circuit that will be used for the design is shown in Figure 1-15a. Our main task, which differs from the design procedure which uses a voltage operational amplifier, is the design of the bias circuit to provide the required open-loop transfer characteristic. We shall demonstrate the design by a step-by-step procedure.

We shall use the CA 3060 amplifier, the characteristics of which as supplied by the manufacturer, are shown at the end of the chapter. We shall make use of the bias regulator supplied inside the I.C. package to minimize the voltage and current offset, depending on supply voltage variations. The use of the bias regulator causes more current drain from the power supply and must be avoided if current drain is critical.

The input offset voltage V_{IO}, the bias current I_B and the input offset current I_{IO} all increase with an increase in the amplifier bias current I_{ABC}. Thus, to keep the total effective offset voltage, $V_{equ. \, offset} = V_{IO} + R_s I_{IO}$, as small as possible, I_{ABC} should be kept as small as possible, large enough only to provide the needed transconductance g_m and sufficient output swing.

Calculation of required g_m. The open-loop voltage gain A_{OL} of the amplifier should be at least 10 times larger than the closed-loop voltage gain A_{CL} to allow for sufficient feedback.

$$A_{OL} \ = \ 10 \ A_{CL} \ = \ 10 \times 6 \ = \ 60$$

But

$$g_m \ = \ \frac{A_{OL}}{R_L} \ = \ \frac{60}{10 \ K\Omega} \ = \ 6 \ \text{milliohms}$$

Calculation of I_{ABC}. From the curve supplied by the manufacturer, we find that for g_m = 6 mmhos, we need I_{ABC} = 30 μA (using the "minimum" curve for g_m). From the I_{OM} curve, we find that the corresponding minimum output current is 60 μA. We shall check below to see if this is sufficient current for our design.

Input Resistor R_S the value of R_S is determined by the specification of the required input resistance of the amplifier.

$$R_S = 25 \ K\Omega$$

Calculation of feedback resistor R_F. The closed-loop gain of an inverting amplifier is given as $A_{CL} \ = \ R_F/R_S$. Hence,

$$R_F = A_{CL} R_S = 6 \times 25 \ K\Omega = 150 \ K\Omega$$

Calculation of R_B. The bias resistor R_B is used to minimize voltage offset resulting from bias currents just as in the voltage operational amplifier.

$$R_B = R_S \| R_F = \frac{R_S R_F}{R_S + R_F} = \frac{25 \text{ K}\Omega \times 150 \text{ K}\Omega}{25 \text{ K}\Omega + 150 \text{ K}\Omega} = 21.4 \text{ K}\Omega \simeq 22\text{K}\Omega$$

Calculation of total output current. The output current is the sum of the current through the load R_L and the feedback resistor R_F. The output voltage v_o is equal to the voltage gain times the signal input voltage v_s.

$$v_o = 6 \times 0.1 \text{ V} = 0.6 \text{ V peak-to-peak}$$

The total current is therefore

$$i_o = i_L + i_F = \frac{v_o}{R_L} + \frac{v_o}{R_F} = \frac{0.6 \text{ V}}{10 \text{ k}\Omega} + \frac{0.6 \text{ V}}{150 \text{ K}\Omega} =$$

$$64 \text{ } \mu\text{A peak-to-peak}$$

This corresponds to a peak current of $64/2 = 32 \text{ } \mu\text{A}$ which is well below the $60 \text{ } \mu\text{A}$ figure obtained from the data sheet for $I_{ABC} = 30 \text{ } \mu\text{A}$.

Calculation of R_Z and R_{ABC}. The bias regulator contains two transistors and a zener diode. To ensure that the bias regulator provides good regulation, the zener diode current should be equal to at least $1.5 I_{ABC} = 1.5 \times 30 \text{ } \mu\text{A} = 45 \text{ } \mu\text{A}$. This ensures that the zener diode remains in its voltage saturation (control) region. The zener diode can be considered to be connected in series with R_Z and the power supplies. The zener voltage is 0.7 V. The value of R_Z is then obtained as follows:

$$1.5 \text{ } I_{ABC} \times R_Z = V^+ + V^- - 0.7 \text{ V}$$

$$R_Z = \frac{6 \text{ V} + (-6 \text{ V}) - 0.7 \text{ V}}{45 \text{ } \mu\text{A}} = 111 \text{ K}\Omega \simeq 110 \text{ K}\Omega$$

The value of R_{ABC} is determined by the required I_{ABC}:

$$R_{ABC} = \frac{V_1 - V_6}{I_{ABC}}$$

where the subscripts 1 and 6 stand for the terminal numbers in Figure 1-15a.

The values of V_1 (bias regulator voltage) and V_2 (amplifier bias voltage) are obtained from the bias regulator characteristics shown in Figure 1-15b. We see that for $I_{ABC} = 30 \text{ } \mu\text{A}$, the curves give $V_1 = 6.8 \text{ V}$ and $V_2 = 0.625 \text{ V}$ giving

$$R_{ABC} = \frac{6.8 - 0.625}{30 \ \mu A} = 206 \ K\Omega \simeq 200 \ K\Omega$$

Summary of design. The circuit is as shown in Figure 1-15a with the following values of components:

$$
\begin{aligned}
R_S &= 25 \ K \\
R_F &= 150 \ K \\
R_B &= 22 \ K \\
R_L &= 10 \ K \\
R_Z &= 110 \ K \\
R_{ABC} &= 200 \ K
\end{aligned}
$$

The 0.1 μF capacitors are used to decouple the circuit from the power supplies.

1.9   PITFALLS TO AVOID

Essentially, to avoid poor designs, the designer must familiarize himself carefully with the operational amplifier specifications and take into consideration those characteristics which may affect critically the design specifications. For example, in the design example of Section 1.8 we saw that certain limits were imposed on the input resistor R_1 by the input offset current. The objective was to ensure that the cumulative percentage distortion error not exceed the specified percentage. The designer must be flexible. Often it is possible to allow a larger contribution to the error by one parameter, while decreasing the contribution by another parameter to obtain an overall compatible system. If voltage regulation or common mode signals contribute potential problems, their effects must be considered as was explained in Sections 1.2.5 and 1.6. In low level amplifier designs, noise becomes an important consideration. Chapter 3 discusses noise problems, and Chapter 11 discusses shielding and proper grounding.

In noninverting amplifiers, unlike inverting amplifiers, the voltage levels of the inverting and noninverting input terminals of the operational amplifier are not kept near ground, but at the signal level. An operational amplifier must be chosen whose common mode range is equal to, or exceeds, the signal magnitude.

Special precautions must be taken with the transconductance amplifier. Both the gain-frequency response and the slew rate are very sensitive to stray as well as load capacitance. This is a result of the high impedances in the circuit. For example, a 10 $K\Omega$ load resistance and an associated stray capacitance of only 15 pF produces a break frequency of 1 MHz as computed from $f = 1/2 \pi RC$. This is shown in Figure 1-15b.

The slew rate $SR = dv_{out}/dt$ is given by the maximum possible output current I_{om} and the capacitance C_L which must be charged: $SR = I_{om}/C_L$. Since the currents in a transconductance amplifier are in the microamp range, even relatively small values of capacitances will cause poor slew rates. For example, if $I_{om} = 10 \mu A$ and $C_L = 100$ pF, the slew rate is only $0.1 \text{ V}/\mu s$.

It must be remembered that because of the high output impedance of the transconductance amplifier, the feedback resistor as well as the load resistor constitute a load on the amplifier.

Samples of operational amplifiers, with their data sheets follow on pages 61 through 74.

Operational Amplifiers

LM747/LM747C dual operational amplifier[1]

general description

The LM747 and the LM747C are general purpose dual operational amplifiers. The two amplifiers share a common bias network and power supply leads. Otherwise, their operation is completely independent.

features

- No frequency compensation required
- Short-circuit protection
- Wide common-mode and differential voltage ranges

- Low-power consumption
- No latch-up
- Balanced offset null

Additional features of the LM747 and LM747C are: no latch-up when input common mode range is exceeded, freedom from oscillations, and package flexibility.

The LM747C is identical to the LM747 except that the LM747C has its specifications guaranteed over the temperature range from 0°C to 70°C instead of -55°C to $+125^\circ$C.

schematic diagram (each amplifier)

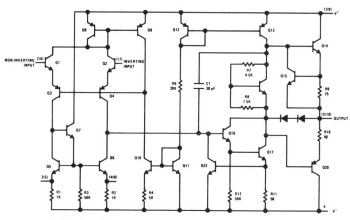

Note: Numbers In Parentheses Are Pin Numbers for Amplifier B. DIP Only.

connection diagrams

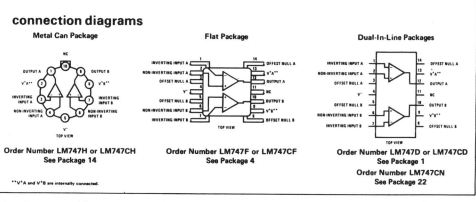

Metal Can Package	Flat Package	Dual-In-Line Packages
Order Number LM747H or LM747CH See Package 14	Order Number LM747F or LM747CF See Package 4	Order Number LM747D or LM747CD See Package 1
		Order Number LM747CN See Package 22

**V⁺A and V⁺B are internally connected.

[1]The 741 amplifier has identical characteristics. (Reprinted by permission of National Semiconductors.

61

absolute maximum ratings

Supply Voltage LM747	±22V
LM747C	±18V
Power Dissipation (Note 1)	800 mW
Differential Input Voltage	±30V
Input Voltage (Note 2)	±15V
Output Short-Circuit Duration	Indefinite
Operating Temperature Range LM747	−55°C to 125°C
LM747C	0°C to 70°C
Storage Temperature Range	−65°C to 150°C
Lead Temperature (Soldering, 10 sec)	300°C

electrical characteristics (Note 3)

PARAMETER	CONDITIONS	LM747			LM747C			UNITS
		MIN	TYP	MAX	MIN	TYP	MAX	
Input Offset Voltage	$T_A = 25°C$, $R_S \leq 10\,k\Omega$		1.0	5.0		1.0	6.0	mV
Input Offset Current	$T_A = 25°C$		80	200		80	200	nA
Input Bias Current	$T_A = 25°C$		200	500		200	500	nA
Input Resistance	$T_A = 25°C$	0.3	1.0		0.3	1.0		$M\Omega$
Supply Current Both Amplifiers	$T_A = 25°C$, $V_S = ±15V$		3.0	5.6		3.0	5.6	mA
Large Signal Voltage Gain	$T_A = 25°C$, $V_S = ±15V$ $V_{OUT} = ±10V$, $R_L \geq 2\,k\Omega$	50	160		50	160		V/mV
Input Offset Voltage	$R_S \leq 10\,k\Omega$			6.0			7.5	mV
Input Offset Current				500			300	nA
Input Bias Current				1.5			0.8	μA
Large Signal Voltage Gain	$V_S = ±15V$, $V_{OUT} = ±10V$ $R_L \geq 2\,k\Omega$	25			25			V/mV
Output Voltage Swing	$V_S = ±15V$, $R_L = 10\,k\Omega$	±12	±14		±12	±14		V
	$R_L = 2\,k\Omega$	±10	±13		±10	±13		V
Input Voltage Range	$V_S = ±15V$	±12			±12			V
Common Mode Rejection Ratio	$R_S \leq 10\,k\Omega$	70	90		70	90		dB
Supply Voltage Rejection Ratio	$R_S \leq 10\,k\Omega$	77	96		77	96		dB

Note 1: The maximum junction temperature of the LM747 is 150°C, while that of the LM747C is 100°C. For operating at elevated temperatures, devices in the TO-5 package must be derated based on a thermal resistance of 150°C/W, junction to ambient, or 45°C/W, junction to case. For the flat package, the derating is based on a thermal resistance of 185°C/W when mounted on a 1/16-inch-thick epoxy glass board with ten, 0.03-inch-wide, 2-ounce copper conductors. The thermal resistance of the dual-in-line package is 100°C/W, junction to ambient.

Note 2: For supply voltages less than ±15V, the absolute maximum input voltage is equal to the supply voltage.

Note 3: These specifications apply for $V_S = ±15V$ and $-55°C \leq T_A \leq 125°C$, unless otherwise specified. With the LM747C, however, all specifications are limited to $0°C \leq T_A \leq 70°C$ $V_S = ±15V$.

typical performance characteristics

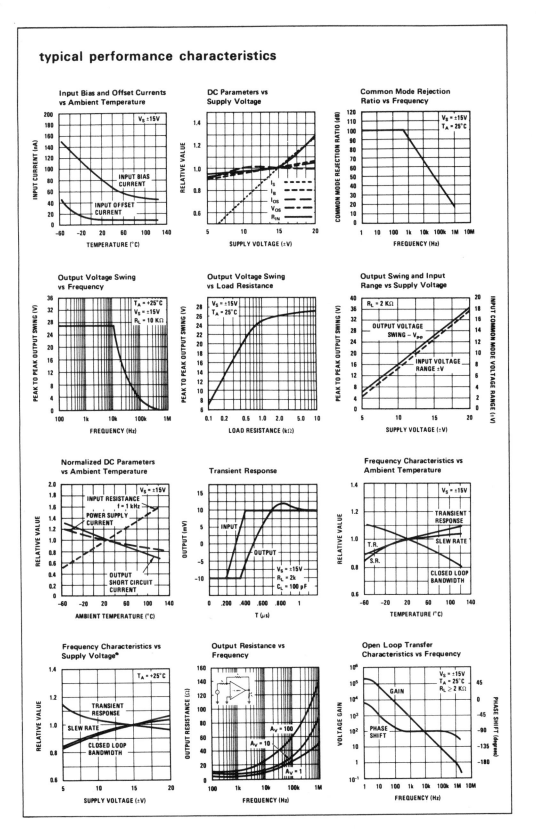

typical performance characteristics (con't)

Input Resistance and Input Capacitance vs Frequency

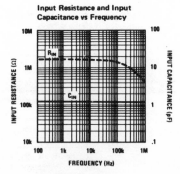

Broadband Noise for Various Bandwidths

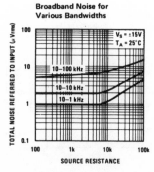

Input Noise Voltage and Current vs Frequency

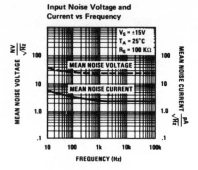

Voltage Follower Large Signal Pulse Response

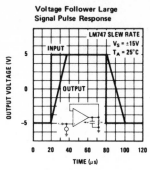

LM748/LM748C operational amplifier

general description

The LM748/LM748C is a general purpose operational amplifier built on a single silicon chip. The resulting close match and tight thermal coupling gives low offsets and temperature drift as well as fast recovery from thermal transients. In addition, the device features:

- Frequency compensation with a single 30 pF capacitor
- Operation from ±5V to ±20V
- Low current drain: 1.8 mA at ±20V
- Continuous short-circuit protection
- Operation as a comparator with differential inputs as high as ±30V

- No latch-up when common mode range is exceeded.
- Same pin configuration as the LM101.

The unity-gain compensation specified makes the circuit stable for all feedback configurations, even with capacitive loads. However, it is possible to optimize compensation for best high frequency performance at any gain. As a comparator, the output can be clamped at any desired level to make it compatible with logic circuits.

The LM748 is specified for operation over the −55°C to +125°C military temperature range. The LM748C is specified for operation over the 0°C to +70°C temperature range.

connection diagrams

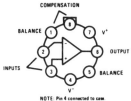

NOTE: Pin 4 connected to case.

Order Number LM748H or LM748CH
See Package 11

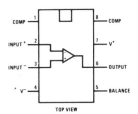

TOP VIEW

Order Number LM748CN
See Package 20

typical applications

Inverting Amplifier with Balancing Circuit

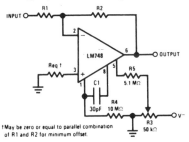

†May be zero or equal to parallel combination of R1 and R2 for minimum offset.

Voltage Comparator for Driving
DTL or TTL Integrated Circuits

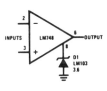

Low Drift Sample and Hold

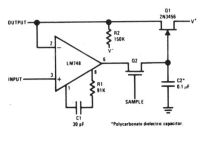

*Polycarbonate dielectric capacitor.

Voltage Comparator for Driving
RTL Logic or High Current Driver

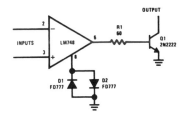

Reprinted by permission of National Semiconductors

65

absolute maximum ratings

Supply Voltage	±22V
Power Dissipation (Note 1)	500 mW
Differential Input Voltage	±30V
Input Voltage (Note 2)	±15V
Output Short-Circuit Duration (Note 3)	Indefinite
Operating Temperature Range: LM748	-55°C to $+125^\circ$C
LM748C	0°C to $+70^\circ$C
Storage Temperature Range	-65°C to $+150^\circ$C
Lead Temperature (Soldering, 10 sec)	300°C

electrical characteristics (Note 4)

PARAMETER	CONDITIONS	MIN	TYP	MAX	UNITS
Input Offset Voltage	$T_A = 25^\circ$C, $R_S \leq 10$ kΩ		1.0	5.0	mV
Input Offset Current	$T_A = 25^\circ$C		40	200	nA
Input Bias Current	$T_A = 25^\circ$C		120	500	nA
Input Resistance	$T_A = 25^\circ$C	300	800		kΩ
Supply Current	$T_A = 25^\circ$C, $V_S = \pm15$V		1.8	2.8	mA
Large Signal Voltage Gain	$T_A = 25^\circ$C, $V_S = \pm15$V $V_{OUT} = \pm10$V, $R_L \geq 2$ kΩ	50	160		V/mV
Input Offset Voltage	$R_S \leq 10$ kΩ			6.0	mV
Average Temperature Coefficient of Input Offset Voltage	$R_S \leq 50\Omega$		3.0		μV/$^\circ$C
	$R_S \leq 10$ kΩ		6.0		μV/$^\circ$C
Input Offset Current	$T_A = 0^\circ$C to 70°C			300	nA
	$T_A = -55^\circ$C to 125°C			500	nA
Input Bias Current	$T_A = 0^\circ$C to 70°C			0.8	μA
	$T_A = -55^\circ$C to 125°C			1.5	μA
Supply Current	$T_A = +125^\circ$C, $V_S = \pm15$V		1.2	2.25	mA
	$T_A = -55^\circ$C to 125°C		1.9	3.3	mA
Large Signal Voltage Gain	$V_S = \pm15$V, $V_{OUT} = \pm10$V $R_L \geq 2$ KΩ	25			V/mV
Output Voltage Swing	$V_S = \pm15$V, $R_L = 10\Omega$	±12	±14		V
	$R_L = 2$ kΩ	±10	±13·		V
Input Voltage Range	$V_S = \pm15$V	±12			V
Common Mode Rejection Ratio	$R_S \leq 10$ kΩ	70	90		dB
Supply Voltage Rejection Ratio	$R_S \leq 10$ kΩ	77	90		dB

Note 1: For operating at elevated temperatures the devices must be derated based on a maximum junction to case thermal resistance of 45°C per watt, or 150°C per watt junction to ambient. (See Curves).

Note 2: For supply voltages less than ±15V, the absolute maximum input voltage is equal to the supply voltage.

Note 3: Continuous short circuit is allowed for case temperatures to +125°C and ambient temperatures to +70°C.

Note 4: These specifications apply for ±5V $\leq V_S \leq$ +15V and -55°C $\leq T_A \leq$ 125°C, unless otherwise specified. With the LM748C, however, all temperature specifications are limited to 0°C $\leq T_A \leq 70^\circ$C.

guaranteed performance characteristics (Note 4)

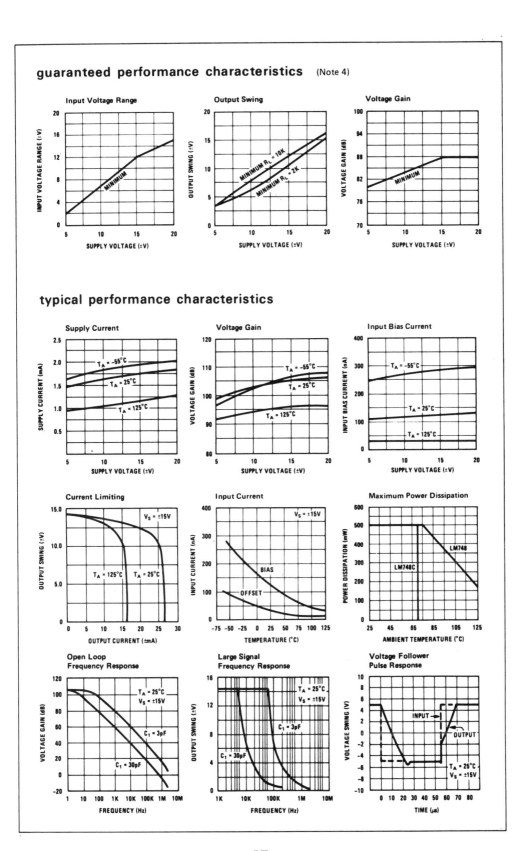

typical performance characteristics

LH0022/LH0022C* high performance FET op amp
LH0042/LH0042C low cost FET op amp
LH0052/LH0052C precision FET op amp

general description

The LH0022/LH0042/LH0052 are a family of FET input operational amplifiers with very closely matched input characteristics, very high input impedance, and ultra-low input currents with no compromise in noise, common mode rejection ratio, open loop gain, or slew rate. The internally laser nulled LH0052 offers 200 microvolts maximum offset and $5\,\mu V/^\circ C$ offset drift. Input offset current is less than 100 femtoamps at room temperature and 100 pA maximum ·at $125^\circ C$. The LH0022 and LH0042 are not internally nulled but offer comparable matching characteristics. All devices in the family are internally compensated and are free of latch-up and unusual oscillation problems. The devices may be offset nulled with a single 10k trimpot with neglible effect in offset drift or CMRR.

The LH0022, LH0042 and LH0052 are specified for operation over the $-55^\circ C$ to $+125^\circ C$ military temperature range. The LH0022C, LH0042C and LH0052C are specified for operation over the $-25^\circ C$ to $+85^\circ C$ temperature range.

features

- Low input offset current – 100 femtoamps max. (LH0052)
- Low input offset drift – $5\,\mu V/^\circ C$ max (LH0052)
- Low input offset voltage – 100 microvolts–typ.
- High open loop gain – 100 dB typ.
- Excellent slew rate – 3.0 V/µs typ.
- Internal 6 dB/octave frequency compensation
- Pin compatible with standard IC op amps (TO-5 package)

The LH0022/LH0042/LH0052 family of IC op amps are intended to fulfill a wide variety of applications for process control, medical instrumentation, and other systems requiring very low input currents and tightly matched input offsets. The LH0052 is particularly suited for long term high accuracy integrators and high accuracy sample and hold buffer amplifiers. The LH0022 and LH0042 provide low cost high performance for such applications as electrometer and photocell amplification, pico-ammeters, and high input impedance buffers.

Special electrical parameter selection and custom built circuits are available on special request.

For additional application information and information on other National operational amplifiers, see *Available Linear Applications Literature*.

schematic and connection diagrams

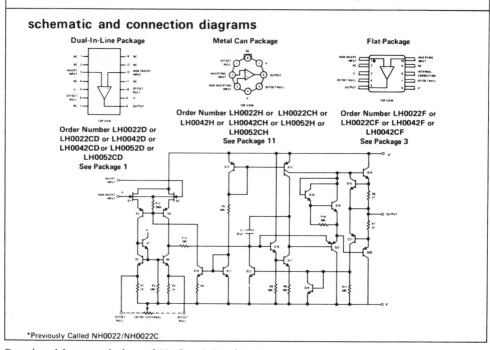

Dual-In-Line Package

Order Number LH0022D or LH0022CD or LH0042D or LH0042CD or LH0052D or LH0052CD
See Package 1

Metal Can Package

TOP VIEW

Order Number LH0022H or LH0022CH or LH0042H or LH0042CH or LH0052H or LH0052CH
See Package 11

Flat-Package

TOP VIEW

Order Number LH0022F or LH0022CF or LH0042F or LH0042CF
See Package 3

*Previously Called NH0022/NH0022C

Reprinted by permission of National Semiconductors

absolute maximum ratings

Supply Voltage	±22V
Power Dissipation (see graph)	500 mW
Input Voltage (Note 1)	±15V
Differential Input Voltage (Note 2)	±30V
Voltage Between Offset Null and V^-	±0.5V
Short Circuit Duration	Continuous
Operating Temperature Range	
LH0022, LH0042, LH0052	-55°C to $+125^\circ$C
LH0022C, LH0042C, LH0052C	-25°C to $+85^\circ$C
Storage Temperature Range	-65°C to $+150^\circ$C
Lead Temperature (Soldering, 10 sec)	300°C

dc electrical characteristics For LH0022/LH0022C (Note 3)

PARAMETER	CONDITIONS	LIMITS						UNITS
		LH0022			LH0022C			
		MIN	TYP	MAX	MIN	TYP	MAX	
Input Offset Voltage	$R_S \leq 100$ kΩ; $T_A = 25^\circ$C		2.0	4.0		3.5	6.0	mV
	$R_S \leq 100$ kΩ			5.0			10.0	mV
Temperature Coefficient of Input Offset Voltage	$R_S \leq 100$ kΩ		5	10		5	15	μV/$^\circ$C
Offset Voltage Drift with Time			3			4		μV/week
Input Offset Current	$T_A = 25^\circ$C		0.2	2.0		1.0	5.0	pA
				200			200	pA
Temperature Coefficient of Input Offset Current		Doubles every 20°C			Doubles every 20°C			
Offset Current Drift with Time			0.1			0.1		pA/week
Input Bias Current	$T_A = 25^\circ$C		5	10		10	25	pA
				1.0			1.0	nA
Temperature Coefficient of Input Bias Current		Doubles every 20°C			Doubles every 20°C			
Differential Input Resistance			10^{12}			10^{12}		Ω
Common Mode Input Resistance			10^{12}			10^{12}		Ω
Input Capacitance			4.0			4.0		pF
Input Voltage Range	$V_S = \pm15$V	±12	±13.5		±12	±13.5		V
Common Mode Rejection Ratio	$R_S \leq 10$ kΩ, $V_{IN} = \pm10$V	80	90		70	90		dB
Supply Voltage Rejection Ratio	$R_S \leq 10$ kΩ, ±5V $\leq V_S \leq \pm15$V	80	90		70	90		dB
Large Signal Voltage Gain	$R_L = 2$ kΩ, $V_{OUT} = \pm10$V, $T_A = 25^\circ$C, $V_S = \pm15$V	100	200		75	160		V/mV
	$R_L = 2$ kΩ, $V_{OUT} = \pm10$V, $V_S = \pm15$V	50			50			V/mV
Output Voltage Swing	$R_L = 1$ kΩ, $T_A = 25^\circ$C, $V_S = \pm15$V	±10	±12.5		±10	±12		V
	$R_L = 2$ kΩ, $V_S = \pm15$V	±10			±10			V
Output Current Swing	$V_{OUT} = \pm10$V, $T_A = 25^\circ$C	±10	±15		±10	±15		mA
Output Resistance			75			75		Ω
Output Short Circuit Current			25			25		mA
Supply Current	$V_S = \pm15$V		2.0	2.5		2.4	2.8	mA
Power Consumption	$V_S = \pm15$V			75			85	mW

dc electrical characteristics for LH0042/LH0042C

($T_A = 25°C$, $V_S = \pm15V$; unless otherwise specified)

PARAMETER	CONDITIONS	LH0042 MIN	LH0042 TYP	LH0042 MAX	LH0042C MIN	LH0042C TYP	LH0042C MAX	UNITS
Input Offset Voltage	$R_S \leq 100\,k\Omega$; $\pm5V \leq V_S \leq 20V$		5.0	20		6.0	20	mV
Temperature Coefficient of Input Offset Voltage	$R_S \leq 100\,k\Omega$		5	20		10	25	µV/°C
Offset Voltage Drift with Time			7			10		µV/week
Input Offset Current			1	5		2	10	pA
Temperature Coefficient of Input Offset Current			Doubles every 20°C			Doubles every 20°C		
Offset Current Drift with Time			0.1			0.1		pA/week
Input Bias Current			10	25		15	50	pA
Temperature Coefficient of Input Bias Current			Doubles every 20°C			Doubles every 20°C		
Differential Input Resistance			10^{12}			10^{12}		Ω
Common Mode Input Resistance			10^{12}			10^{12}		Ω
Input Capacitance			4.0			4.0		pF
Input Voltage Range		±12	±13.5		±12	±13.5		V
Common Mode Rejection Ratio	$R_S \leq 10\,k\Omega$, $V_{IN} = \pm10V$	70	86		70	80		dB
Supply Voltage Rejection Ratio	$R_S \leq 10\,k\Omega$, $\pm5V \leq V_S \leq 15V$	70	86		70	80		dB
Large Signal Voltage Gain	$R_L = 1\,k\Omega$, $V_{OUT} = \pm10V$	50	150		25	100		V/mV
Output Voltage Swing	$R_L = 1\,k\Omega$	±10	±12.5		±10	±12		V
Output Current Swing	$V_{OUT} = \pm10V$	±10	±15		±10	±15		mA
Output Resistance			75			75		Ω
Output Short Circuit Current			20			20		mA
Supply Current			2.5	3.5		2.8	4.0	mA
Power Consumption				105			120	mW

dc electrical characteristics For LH0052/LH0052C (Note 3)

PARAMETER	CONDITIONS	LH0052 MIN	LH0052 TYP	LH0052 MAX	LH0052C MIN	LH0052C TYP	LH0052C MAX	UNITS
Input Offset Voltage	$R_S \leq 100\,k\Omega$; $V_S = \pm15V$, $T_A = 25°C$		0.1	0.5		0.2	1.0	mV
	$R_S \leq 100\,k\Omega$, $V_S = \pm15V$			1.0			1.5	mV
Temperature Coefficient of Input Offset Voltage	$R_S \leq 100\,k\Omega$		2	5		5	10	µV/°C
Offset Voltage Drift with Time			2			4		µV/week
Input Offset Current	$T_A = 25°C$		0.01	0.1		0.02	0.2	pA
				100			100	pA
Temperature Coefficient of Input Offset Current			Doubles every 20°C			Doubles every 20°C		
Offset Current Drift with Time			<0.1			<0.1		pA/week
Input Bias Current	$T_A = 25°C$		0.5	1.0		1.0	5.0	pA
				500			500	pA
Temperature Coefficient of Input Bias Current			Doubles every 20°C			Doubles every 20°C		
Differential Input Resistance			10^{12}			10^{12}		Ω
Common Mode Input Resistance			10^{12}			10^{12}		Ω
Input Capacitance			4.0			4.0		pF
Input Voltage Range	$V_S = \pm15V$	±12	±13.5		±12	±13.5		V
Common Mode Rejection Ratio	$R_S \leq 10\,k\Omega$, $V_{IN} = \pm10V$	80	90		76	90		dB
Supply Voltage Rejection Ratio	$R_S \leq 10\,k\Omega$, $\pm5V \leq V_S \leq \pm15V$	80	90		76	90		dB
Large Signal Voltage Gain	$R_L = 2\,k\Omega$, $V_{OUT} = \pm10V$, $V_S = \pm15V$, $T_A = 25°C$	100	200		75	160		V/mV
	$R_L = 2\,k\Omega$, $V_{OUT} = \pm10V$, $V_S = \pm15V$	50			50			V/mV
Output Voltage Swing	$R_L = 1\,k\Omega$, $T_A = 25°C$, $V_S = \pm15V$	±10	±12.5		±10	±12		V
	$R_L = 2\,k\Omega$, $V_S = \pm15V$	±10			±10			V
Output Current Swing	$V_{OUT} = \pm10V$, $T_A = 25°C$	±10	±15		±10	±15		mA
Output Resistance			75			75		Ω
Output Short Circuit Current			25			25		mA
Supply Current	$V_S = \pm15V$		2.0	2.5		2.5	3.0	mA
Power Consumption	$V_S = \pm15V$			75			90	mW

ac electrical characteristics For all amplifiers (T_A = 25°C, V_S = ±15V)

PARAMETER	CONDITIONS	LIMITS						UNITS
		LH0022/42/52			LH0022C/42C/52C			
		MIN	TYP	MAX	MIN	TYP	MAX	
Slew Rate	Voltage Follower	1.5	3.0		1.0	3.0		V/μs
Large Signal Bandwidth	Voltage Follower		40			40		kHz
Small Signal Bandwidth			1.0			1.0		MHz
Rise Time			0.3	1.5		0.3	1.5	μs
Overshoot			10	30		15	40	%
Settling Time (0.1 %)	ΔV_{IN} = 10V		4.5			4.5		μs
Overload Recovery			4.0			4.0		μs
Input Noise Voltage	R_S = 10 kΩ, f_o = 10 Hz		150			150		nV/$\sqrt{Hz}$
Input Noise Voltage	R_S = 10 kΩ, f_o = 100 Hz		55			55		nV/$\sqrt{Hz}$
Input Noise Voltage	R_S = 10 kΩ, f_o = 1 kHz		35			35		nV/$\sqrt{Hz}$
Input Noise Voltage	R_S = 10 kΩ, f_o = 10 kHz		30			30		nV/$\sqrt{Hz}$
Input Noise Voltage	BW = 10 Hz to 10 kHz, R_S = 10 kΩ		12			12		μVrms
Input Noise Current	BW = 10 Hz to 10 kHz		<.1			<.1		pArms

typical performance characteristics

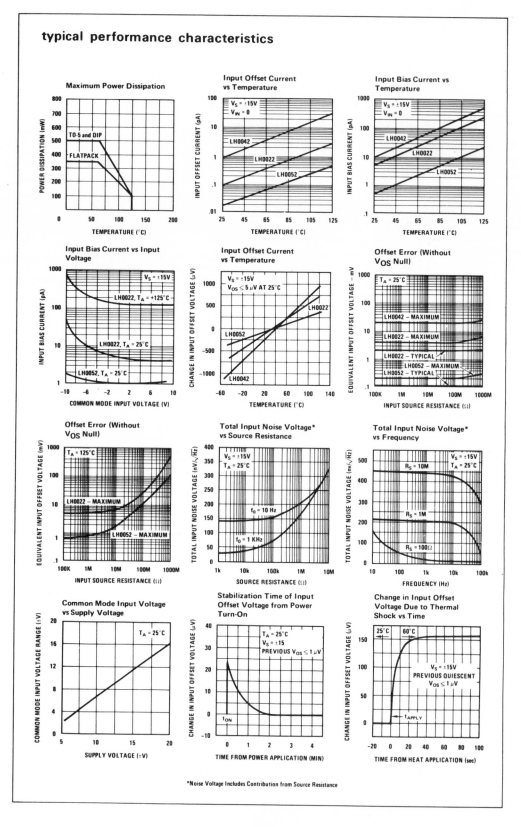

*Noise Voltage Includes Contribution from Source Resistance

typical performance characteristics (con't)

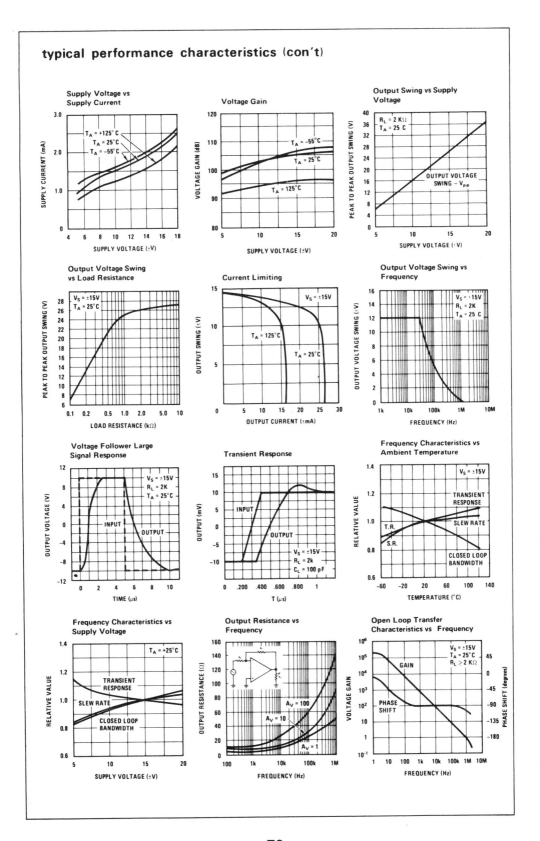

73

Operational Transconductance Amplifier

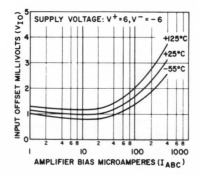

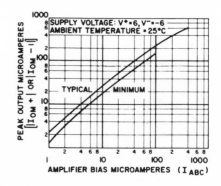

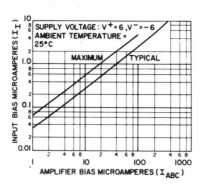

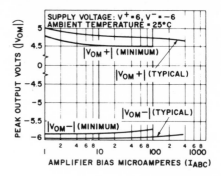

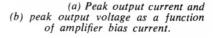

(a) Input offset voltage and
(b) input bias current as a function of
amplifier bias current.

(a) Peak output current and
(b) peak output voltage as a function
of amplifier bias current.

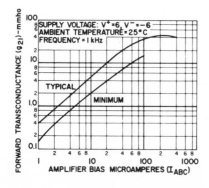

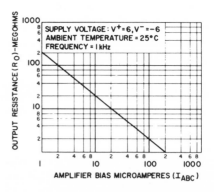

Forward transconductance
as a function of amplifier bias current.

Output resistance as a function of amplifier bias current.

Reprinted by permission of RCA Solid State Division

Chapter 2

Selecting Passive Components

2.1   To complete the design of an electronic circuit, it is necessary to select and specify passive components which include resistors, capacitors, inductors and transformers. In modern circuits, inductors and tranformers are much less common than resistors and capacitors. These components must be assembled together with the I.C.'s onto a common board. The printed circuit board is the most reliable and economical means for the assembly of the final circuit. In this chapter we discuss the various types of available components, the differences between similar components, and how to choose the right components for a given application.

2.2   RESISTORS

Resistors can be divided into two main classes: (1) conventional bulk components, and (2) thick and thin film resistor arrays. Bulk components provide the largest variety, are stocked in large quantities by local suppliers of electronic components, and are the most convenient to use in developmental designs and small quantity production. Thick and thin film resistor arrays have been made available for general use in some variety since the early 1970's. They have the advantage of being contained in standard *DIP* (dual in line) and *SIP* (single in line) packages, making their assembly into a circuit easy and giving the completed circuit the appearance of an assembly of packages rather than individual components. This makes for simple assembly and should be considered for larger production quantities.

2.2.1 Bulk Resistors

There are four major types of bulk resistors. Each type has unique advantages and certain limitations.

Carbon-Composition Resistors

Carbon-Composition resistors are very reliable and they are the least expensive of all resistor types. Consequently, they are the most widely used in electronic circuits. They are available in the range from 1 Ω to 22 MΩ with tolerances of 3%, 5%, 10% and 20%, and power ratings of ⅛ watt, ¼ watt, ½ watt, 1 watt and 2 watts. The temperature coefficient (TC) of carbon-composition resistors is relatively high and can be considered to be an average of 0.1%/°C (or 1000 ppm/°C) over the applicability range up to a temperature of about 160°C. However, in the normal range of application between 0° and 60° C the coefficient is considerably less. It increases rapidly below 0°C and above 60°C. The noise level of carbon-composition resistors is the highest among all resistors.

Wirewound Resistors

Wirewound resistors are divided into three broad categories: high-power, high-accuracy, and general-purpose resistors. Since these resistors are made of wire coils, their inductance is relatively high and must be taken into account at high frequencies. *Bifilar* wound resistors, however, are available at a higher cost to minimize the inductive effect. They are usually referred to as *non-inductive wirewound resistors.*

High-power wirewound resistors are generally available within the 1 Ω to 100 KΩ resistance range with 5-200 watt dissipation, and tolerances of 5% and 10%. Although integrated circuits generally operate at very low power levels, power resistors must be used, for example, in the output stage of an audio amplifier in conjunction with power transistors which are, in turn, fed from an integrated low power amplifier stage.

High-accuracy wirewound resistors exhibit a low temperature coefficient and excellent long-term stability. Most commonly used are 1% resistors, but resistors with *tolerances* as low as 5 ppm are available. Resistors with *temperature coefficients* as low as 5 ppm are also available. Most commonly used power ratings are ¼, ½ and 1-watt, but 3, 5, 10, 25, and 50-watt resistors are also available, and resistance values are in the range from 1 Ω to 100 KΩ. High accuracy resistors are used in applications such as precision active full wave rectifiers or other circuits where precision voltage divider networks are needed.

General-purpose wirewound resistors are available in the ¼ Ω to 10 KΩ resistance range and power ratings of ½, 1, and 3 watts. They duplicate, to some degree, application areas of carbon-composition resistors and are used mainly when low temperature coefficients are needed.

Metal-Film Resistors

Metal-film resistors are suitable for the most demanding applications. Their resistance range spans the gamut from 0.1 Ω to 1.5 MΩ. Resistors with temperature coefficients as low as 1 ppm/°C are available with excellent long-term stability and power ratings of $\frac{1}{10}$, $\frac{1}{8}$, $\frac{1}{4}$, $\frac{1}{2}$, and 1 watt. These resistors have extremely low noise-figures which make them desirable for low-level preamplifiers and other low noise applications.

Carbon-Film Resistors

Two unique characteristics of *carbon-film* resistors are the high available resistance values (10 Ω to 100 MΩ) and the negative temperature coefficient of resistance. The resistance tolerances are 0.5% and higher. They have low noise in resistance values below 100 KΩ and are inexpensive. These resistors are used where high values are needed, as well as for temperature compensation in some circuits where advantage is taken of the negative temperature coefficient.

2.2.2   Thin and Thick Film Resistor Arrays

The era of integrated circuits gave rise to the use of packaged resistor arrays. These are assemblies of individual flat-shaped *DIP* packages with 14, or other specified number, of leads for external connections. By interconnecting external leads, any number of resistors can be connected in series or in parallel to obtain a variety of values. For information on available values the reader should contact the manufacturers. Many are listed in the Electronic Engineers Master (EEM) publication.

The main advantages of these resistor packages are their compactness and neatness and ease of assembly into an electronic system. They should be considered particularly in large quantity runs, where cost savings may be realized.

2.2.3   Pitfalls to Avoid

Nominal resistor values have been chosen and standardized by manufacturers to avoid waste in manufacturing. That is, *any* resistor manufactured subject to process control limitations falls within the tolerances of a nominal value. Accordingly, resistors with ±20% tolerance are available with the following nominal values: 10, 15, 22, 33, 47, 68, 100 ohms and in multiples of 10 of these values; resistors with ±10% tolerance are available with the nominal values of 10, 12, 15, 18, 22, 27, 33, 39, 47, 56, 68, 82, 100 ohms and multiples of 10 of these values; and ±5% tolerance resistors are available with the nominal values of 10, 11, 12, 13, 15, 16, 18, 20, 22, 24, 27, 30, 33, 36, 39, 43, 47, 51, 62, 68, 75, 82, 91, 100 and multiples of 10 of these values. Tighter tolerance resistors are often specified in fractions of ohms and manufacturers' or suppliers' catalogs should be consulted.

In designing circuits, the designer should specify *nominal* resistor values to avoid delay or confusion or poor choice of resistors by the technician assembling the circuit.

Power dissipation of a resistor of value R is given by I^2R or V^2/R where I is the current through the resistor, and V is the voltage across the resistor. It must be noted that resistor power ratings are given for stated ambient temperatures. If no value for the ambient temperature is stated, it is normally safe to assume that the power rating is valued up to about 70° C.

The power rating is limited by the maximum allowable temperature of the resistor. Too high a temperature can destroy a resistor, but even more troublesome to the maintenance person is a permanent change in the resistance value. Too high a temperature can cause an irreversible change of resistance resulting from a change in the resistor material composition. This can cause, for example, a permanent change in amplifier gain or oscillator frequency.

To avoid such problems it is necessary to *derate* the power rating of a resistor when it is to be used at higher temperatures. If no curves or other data for derating are avialable, it is usually reasonable to derate the power rating of the resistor linearly from 100% to 0% rating between 70° C and 150° C according to the formula:

$$\text{Power Rating at } t° C = 100\% \text{ Power Rating } (1 - \frac{t-70}{80})$$

Thus, for example, the power dissipated in a ½-watt resistor should not exceed 375 milliwatts if the circuit is to be operated at an ambient temperature of 90°C.

It must also be borne in mind that resistors are *"noisier"* at higher temperatures. The thermal noise in a resistor results from the random motion of electrons in the resistor and increases with temperature. The thermal noise manifests itself in a voltage across, and a corresponding current through, the resistor. This voltage increases in proportion to the square root of the absolute temperature and is in proportion to the square root of the resistance. The absolute temperature is obtained by adding 273 degrees to the temperature in degrees Celsius. The noise voltage contains all frequencies. It is thus possible to reduce the circuit noise level by filtering the output, allowing only frequencies in the signal frequency range to reach the output.

The temperature coefficient of resistance must be considered at high ambient temperatures or at high power levels when high component temperatures may be reached. This is, however, of no major concern in most designs using IC's, except for output stages, since the power levels are normally very low, considerably below the resistor power ratings. It must also be noted that the change in system performance as a result of a change in resistance values is less than the change in resistance if the performance depends on resistance *ratios*. If, for example, the gain of an amplifier is given by the ratio $R_2/R_1 = 100K/5K = 20$, the TC is $1\%/°C$ and the temperature rise is $20°$ C, then the gain at the higher temperature is $120K/6K = 20$, so that no change in gain has taken place. On the other hand, if the 5K resistor is connected to the input terminal of an operational amplifier, then the voltage drop across the resistor caused by the bias current will change as a result of the resistance change. Unless this voltage drop is very small in comparison with the voltage drop caused by the signal, a change in amplifier performance will take place.

If the performance of a circuit depends on an RC time constant, the performance will, in general, change as a function of temperature since both R and C are temperature dependent. If the circuit must operate over a wide temperature range this can be of concern. It is then necessary to choose components with small TC's. It is interesting to note that high degrees of temperature stability can be obtained by proper choices of resistors and capacitors with TC's of nearly equal magnitudes but opposite signs. For example, a carbon film resistor in combination with a low loss ceramic capacitor can be used. Both components have TC's of the order of 100 ppm, the coefficient being negative for the resistor and positive for the capacitor. Some degree of compensation for resistive networks can be obtained by connecting a carbon-composition resistor with a positive TC in series with a carbon film resistor with a negative TC.

A word of caution about *high voltage limitations* is necessary. The voltage levels usually encountered in I.C. circuit design are of the order of 5, 15, or 30 volts and the voltage, except as it relates to power dissipation, is no cause for concern. However, an I.C. circuit may be used for example in the feedback circuit of a high voltage power supply to stabilize the voltage, and the designer may have to assume the responsibility for the complete system. The high voltage of hundreds or thousands of volts may cause an irreversible change or the destruction of a resistor resulting from the high electric field before the maximum power dissipation is reached. It is, therefore, necessary to consider the voltage rating of a resistor in addition to the power rating. A typical value for wirewound resistors is 1000 V. When unusually high voltages are encountered, the manufacturer's catalog must be consulted.

High frequency effects on resistors must be considered in applications above 200 kHz. Care must be exercised since the effective value of a resistor can be off by several times the nominal D-C value. Wirewound resistors, except the more expensive bifilar resistors, must be avoided because of the inductive effect which adds impedance to the total impedance of the circuit at higher frequencies. But even carbon-composition resistors deviate from their low frequency value at higher frequencies because of the shunt stray capacitance which lowers the impedance. Typical frequency characteristics of carbon-composition resistors are shown in Figure 2-1.

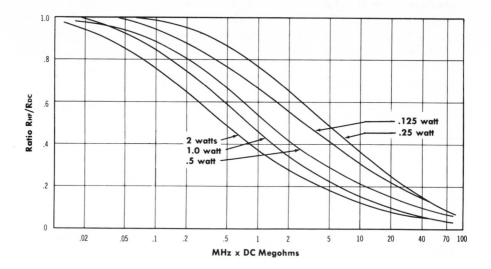

Figure 2-1. Typical high frequency characteristics of carbon composition resistors.

High resistance valued resistors above $100M\Omega$ can be a source of trouble if not handled correctly. The leakage shunt resistance between the terminals lowers the effective resistance particularly at high relative humidity, or in the presence of contaminants on the resistor surface. High valued resistors must be cleaned in a solvent before they are connected into a circuit.

2.2.4   How to Choose the Right Type Resistor

The general rule is: choose the least expensive resistor that will perform the required function satisfactorily. The price of the resistor will usually increase with tighter tolerance and T.C. specifications. In a large variety of applications carbon-composition resistors will meet the requirements. For highest precision, wirewound resistors must be used, and in high frequencies where the inductance is objectional, the bifilar wirewound resistors must be used. Very high values can be obtained only in carbon film resistors and very low values are available in metal film resistors.

2.2.5 Potentiometers (Variable Resistors)

Miniature Potentiometers

The miniaturization of electronic circuits stimulated the development of a large variey of *miniature* and so-called *subminiature* potentiometers. They are convenient to use and have become the standard component, except for front panel mounting.

The potentiometers can be divided into three types according to the material of the resistive element: carbon, cermet, and wirewound potentiometers. In addition to resistance values and power ratings, specified parameters include resistance tolerance and minimum resistance (which are of no importance except when the potentiometer is used at one of the extreme positions, which would indicate a poor choice of potentiometer, and is not the normal case), linearity (which is also of no great importance since the miniature potentiometers are set once for the proper value and are not controlled by the equipment user), resolution, and temperature coefficient of resistance.

The carbon-type potentiometer is least expensive, but it has the widest tolerances and the largest TC. Next in price comes the cermet type, and finally, the wirewound type. It is the most expensive type but it has the tightest specifications. Miniature potentiometers come in rectangular and circular shapes and a variety of ratings. Terminals are in the form of insulated stranded leads, solder lugs, or printed circuit pins. Especially convenient for circuits using IC's, are Dual-In-Line potentiometers which have the standard DIP size (TO-116) compatible with automatic insertion equipment. Electronic suppliers' catalogs list many types and further details can be obtained from manufacturers' data sheets.

Carbon types are available in the range of 5 Ω through 1 MΩ, cermet types in the range of 10 Ω through 100K Ω. Power ratings range from 0.2 to 1 watt. Temperature coefficients can be as low as 100 ppm/$^\circ$C, resistance tolerance as low as $\pm$ 3% and linearity $\pm$ 0.15%. High resolutions are achievable with multiturn potentiometers. 25, 20, 15, 10 and 4-turn pots are available; they cost several times more than single-turn pots which can be used for less demanding circuits. They all have screwdriver adjustments. Normally, these miniature pots are mounted directly on the assembly board, although panel-mount units with heavier screw adjustable shafts and mounting nuts are available for most types at a higher cost.

Examples of typical potentiometer specifications are: 1) wirewound 0.5 watt at 70° C, 25-turn adjustment, absolute minimum resistance 0.7% or 0.5 ohm, whichever is greater, for 10 ohms to 50K; 0-5% for 100 K. Maximum operating temperature 125° C, resistance tolerance 10%, sealed against sand, dust and humidity. Size: 5/16″ × 1/4″ × 1/4″; 2) Cermet 0.75 watt at 25° C; 0 watt at 125° C, infinite resolution; stability ± 0.05%; resistance tolerance ±10%, operating temperature range –55 to +125°C; size: 0.350″ × 0.19″ × 0.75″.

Panel-Mount Potentiometers

Most of the microminiature potentiometers described in the previous section are also availabe for panel mounting, but for ruggedness and for convenience for operator's control, larger size potentiometers are usually used for panel mounting. Most common are 3/4″ diameter 2-watt pots rated at 70° C, with 1/4″ dia. shaft for either knob or screwdriver control.

Panel potentiometers are available with carbon-composition, cermet, and wirewound elements. For general purpose, single-turn (270 degrees rotation) pots are available and for precision control, 5-turn or 10-turn pots are used. Readily available carbon element potentiometers are in the range from 100 ohms to 1 MΩ, cermet elements are in the range 50 Ω to 5 MΩ and wirewound elements in the range 10 Ω to 300 KΩ. The power ratings range from 1/2 watt to 50 watt.

Examples are: 1)hot molded carbon element, 1/2″ diameter, 1/2 watt at 70° C, operating temperature – 55° C to + 120° C, resistance tolerance ± 10%, 1/8″ diameter. 2) Cermet element, 1/2″ diameter, 1.0 watt at 125° C, resistance tolerance ± 10%, 1/8″ diameter shaft. 3) Precision 10-turn pot, 1″ diameter, 2.0 watt at 70° C, maximum operating temperature 125° C, resistance tolerance ± 3%, linearity ± 0.2%, 1/4″ shaft diameter. Count dials, digital readout and clock-face readout dials for easy control setting are also available.

Taper refers to the relation between the fraction of the resistance to the shaft position. In a linear taper pot the resistance is proportional to the shaft position. Other tapers are logarithmic taper used commonly in the audio output stages to conform to the logarithmic sensitivity of the ear to sound amplitude. Sinusoidal and other special tapers are also available in right-handed or left-handed configurations.

Pitfalls to Avoid

Note that the *power rating* derates with temperature. The rating is specified for a given temperature derating to zero power at a given higher temperature. If no graphical data is given, a linear derating curve between the two temperatures can be assumed.

Most important is to note that the power dissipation is given by I^2R where I is the current and R is the resistance. The rating is specified for the *total resistance* of the potentiometer. If only a fraction of the resistance is used in the circuit, the permissible power dissipation must be reduced to the same fraction of the specified power rating.

The term *infinite resolution* can be misleading when applied to practice. Carbon composition and cermet element pots have infinite resolution since, in contrast with wirewound elements, the elements are continuous in the direction of wiper movement. However if the potentiometer is a one-turn pot, the resistance depends very critically on the exact position of the wiper, and settings to an exact value are very difficult to obtain. On the other hand, in multi-turn pots the resistance changes relatively little with shaft position, and needed resistance value can be obtained relatively easily, even if the element is wirewound and has finite resolution.

A high degree of *linearity* is important only if the resistance value (or corresponding gain, etc.) is set by a dial reading. If, on the other hand, the value is set by a voltage reading, the extras cost for better linearity is not justified.

2.3 CAPACITORS

Selecting the "right" capacitor from among the great variety of available types can be a confusing task for the circuit designer. Capacitors are characterized by voltage rating, insulation resistance, dissipation factor, frequency response, size and, of course, capacitance. In addition, the temperature dependence of several of these parameters can be quite large. For a particular application we must weigh the importance of each factor and select the capacitor which most closely meets our needs. A casual selection may result in degraded circuit performance and over-specification will certainly result in increased cost and size.

A capacitor with one or two high quality specifications can cost more than many I.C.'s and will probably be larger! With the exception of the I.C.'s, few circuit components are required as much attention to detail in their specification as does the capacitor.

2.3.1 Capacitor Specifications

Each capacitor type (ceramic, film, electrolytic, etc.) has sufficiently different characteristics and areas of applicability to warrant individual discussion. But first let us examine and clarify some of the specifications applied to the capacitor.

Capacitance

The specification seems almost trivial but an understanding of what gives a capacitor its capacitance will aid in understanding many other characteristics. All capacitors consist basically of two metal electrodes separated by an insulator which is called the dielectric (See Figure 2-2). In the ceramic capacitor, for example, this structure is realized by depositing a metal film on both sides of a fired clay disc and attaching leads. A metalized polycarbonate capacitor consists of two plastic films with metal deposited on one side of each, laid on top of each other, and wound in a spiral to form a tubular shape. In any case, Figure 2-2 is an appropriate mechanical representation. The capacitance C of this structure is given by

$$C = \frac{\epsilon_0 \, KA}{d} \tag{2-1}$$

where ϵ_0 = permittivity of free space, A = area of either plate, K = dielectric constant of the dielectric material, and d = distance between the plates.

To increase the capacitance for a given size we could reduce d and by using more layers, increase A. But this process is limited by the maximum field strength rating of the dielectric since, for a given voltage, the electrical field increases as we decrease the thickness.

We have no control over ϵ_0, but K, the dielectric constant, can range from 1 to several thousand for various insulating materials and is, therefore, an important parameter. Almost all capacitance variations with temperature, frequency and applied voltage are due to variations in K.

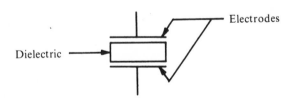

Dielectric

Electrodes

Figure 2-2. Schematic diagram of capacitor.

Insulation Resistance (D.C. Leakage)

This specification is a measure of the capacitor's ability to block d.c. currents. It is defined by measuring the current I_L flowing through the capacitor when a d.c. voltage V_{dc} is applied across the terminals and calculating the equivalent resistance $R_s = V_{dc}/I_L$. The effect of this resistance can be modeled by the circuit of Figure 2-3. Looking at Equation 2-1, with ϵ_o, K and d fixed, the capacitance increases with increased surface area A and, since the leakage flows in the insulating material, the leakage current increases proportionally and the equivalent shunt resistance R_S decreases. For a given type capacitor, C, and given voltage rating, we find that the product $R_S \times C$ is constant. This number is called the "megohm-microfarad product" and, since its units are seconds, it is the time constant of the capacitor C and its shunt resistance R_S. Typical high quality film capacitors have time constants, $R_S \times C$ in excess of 50,000 seconds (14 hours). At low capacitance values, this relationship implies extremely high R_S values and in practice manufacturers specify $R_S \times C$ and a maximum value of R_S which need not be exceeded.

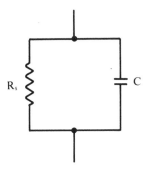

Figure 2-3. Equivalent circuit of capacitor considering insulation resistance.

For long-term integrators and sample hold applications, the insulation resistance is the important specification.

Example. Find the equivalent parallel resistance R_S of a $1\mu f$ polycarbonate capacitor whose megohm-microfarad product is 50,000 seconds.

Solution:

$$C \times R_S = 50,000 \text{ seconds, so}$$

$$R_S = \frac{50,000 \text{ sec}}{1 \ \mu\text{farad}} = 50,000 \text{ M}\Omega.$$

Electrolytic capacitors have a voltage sensitive insulation resistance and normally manufacturers specify the actual d.c. leakage current at the rated voltage instead of the equivalent resistance.

Dissipation Factor, D.F., Quality Factor, Q, and
Equivalent Series Resistance, ESR

The dissipation factor is a measure of the power losses in a capacitor under a.c. conditions. In a few capacitors this loss is partially composed of $I^2 \times R$ losses in the electrodes, but for most capacitors, all the a.c. losses occur in the dielectric material. This effect results from the microscopic motion and resultant heating of small elements of the dielectric caused by the varying forces associated with the alternating electric field. The dissipation constant is defined as the total energy lost in the capacitor, divided by the total energy stored in the capacitor, where the exciting wave form is a single cycle of a sine wave. Dissipation factor is normally expressed as a percentage of stored energy. For a general purpose ceramic capacitor, the D.F. is approximately 3%. For high quality capacitors, typical values are on the order of tenths of one percent. It is important to note that the D.F. is the loss per cycle and even though the D.F. is relatively constant over a broad frequency range for many materials, the total power loss increases linearly with frequency. Thus, for a capacitor C used in a sine wave application at frequency f, we have:

$$\text{Power loss} = \tfrac{1}{2} C(v)^2 \times \text{D.F.} \times f$$
where v = rms value of the sine wave.

The Q factor is the inverse of the dissipation factor: $Q = 1/\text{D.F.}$ If D.F. = .01, Q = 100, and if D.F. = 0.1% then Q = 10,000. The D.F. or Q factors are of interest when designing high Q resonant tanks and other a.c. applications.

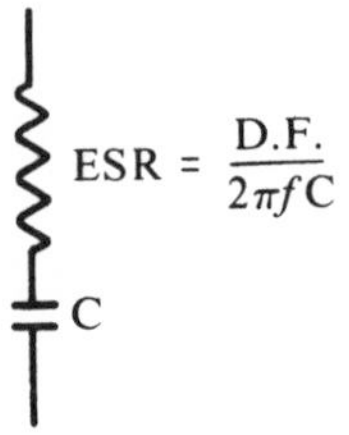

$$\text{ESR} = \frac{\text{D.F.}}{2\pi f C}$$

Figure 2-4. Equivalent circuit of capacitor considering the dissipation factor.

We can model these losses by placing an equivalent resistance in series with the capacitor as in Figure 2-4. However, since D.F. is relatively constant the Equivalent Series Resistance, ESR, is frequency dependent. Specifically:

$$\text{ESR} = \frac{\text{D.F.}}{2\pi f\ C} \text{ where f = the operating frequency.}$$

Since the power loss in electrolytic capacitors is a combination of electrode resistance and dielectric losses, most manufacturers publish a graph of ESR as a function of frequency as the D.F. is not constant in these types.

2.3.2   Types of Capacitors

With the above background, let us now discuss some major families of capacitors commonly used in low voltage level electronic circuits.

Ceramic

Surely, the ceramic capacitor family has the widest range of applicability of all capacitor types. Suppliers have divided this family into three groups based on the dielectric constant K of the ceramic material.

Low K—Temperature compensating and NPO capacitors. These capacitors are fabricated with low K material and are the high performance members of this family. Unfortunately, the low K results in large size for a given voltage rating and capacitance, so these types are limited to about 5000 pF. These types are available in a zero temperature coefficient of capacitance, NPO, and several negative temperature coefficients, N220, N750, etc. The curves of Figure 2-5 demonstrate the excellent performance of this group. Most notable is the stability with parameter variation and high Q factors. These capacitors are excellent choices in r.f. and precision applications at low capacitance values.

Medium K—Temperature stable. This is an intermediate formulation and is frequently advertised as temperature stable although looking at Figure 2-6, we see that the performance is significantly worse than the low K types in all areas. Of course, the higher K value results in more capacitance for a given size. Typical designation of medium K capacitors are X7F, X7R, and Y5R.

High K—General purpose. This group is advertised as "General Purpose for Non-Critical Applications." Looking at Figure 2-7 we see that the performance of this group can only be described as poor. The advantage, of course, is very high capacitance in a small package, but if we are working near the voltage rating or at a temperature extreme, we may not have as much capacitance as we hoped for.

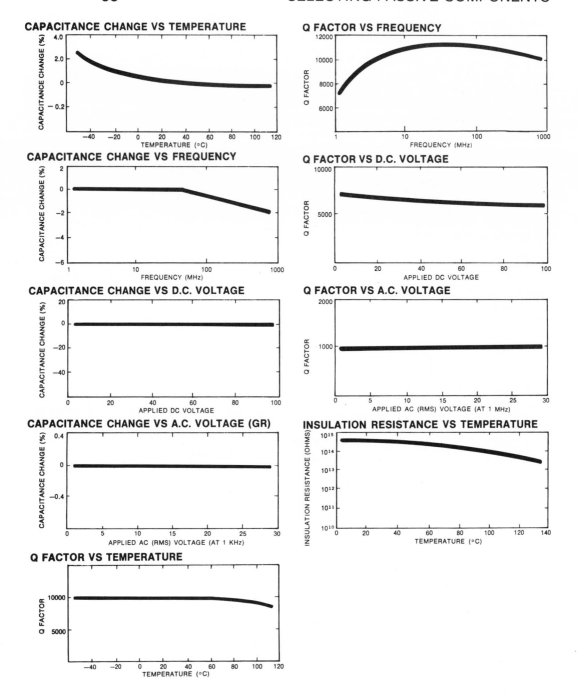

Figure 2-5. Typical performance characteristics of low-K NPO capacitors. (Reprinted with permission of Murata Corporation of America)

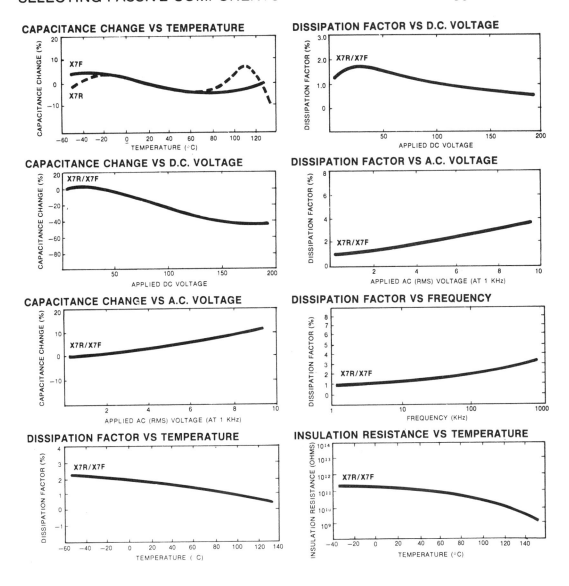

Figure 2-6. Typical performance characteristics of medium-K capacitors.
(Reprinted with permission of Murata Corporation of America)

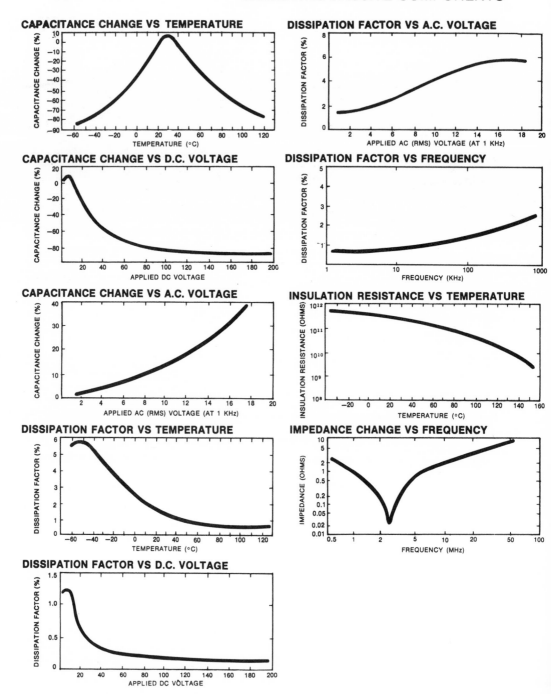

Figure 2-7. Typical performance characteristics of high-K capacitors.
(Reprinted with permission of Murata Corporation of America)

Film Types

The dielectric in this family is a plastic film. The electrodes can be sheets of aluminum or evaporated aluminum deposited directly on the film in a high vacuum (metalized). Typically, tapes of film and metal are alternated and wound in a tubular shape. Depending on the electrode terminating techniques, this construction can exhibit rather high or very low series inductance. Table 2.1 provides a very good performance comparison among the various film types.

Polyester (Mylar*). Except for working voltage, this type has somewhat poorer specifications at 25 ° C than other members of this group. Notable, however, is its wide useful temperature range. Typical performance characteristics are shown in Figure 2-8.

Polystyrene. For low voltage applications and operating temperatures below 100° C, polystyrene is an excellent dielectric. Unfortunately, the low dielectric constant and low working voltage imply that for the same capacitance and voltage rating, the polystyrene capacitor is much larger than other film capacitors. This type has a predictable (–) 120 ppm.°C temperature coefficient. Extremely high Q factors at 25°C over a broad frequency range recommend this type for high Q resonant tanks and filters.

Polycarbonate. This type offers good performance over a wide temperature range.

Polypropylene. At the expense of capacitance stability with temperature, polypropylene offers moderate temperature range performance similar to polystyrene in a somewhat smaller package.

Teflon. Teflon is a very high insulation resistance material with otherwise moderate performance to 200° C.

Mica Capacitors

Mica capacitors out-perform the low K ceramics at very high (> 200 MHz) frequencies. Unfortunately, mica's small dielectric constant results in capacitors with low capacitance to volume ratios. The temperature coefficient can be controlled very closely by a proper choice of the type of mica and the method of construction. It is possible to manufacture mica capacitors with predictable temperature coefficients in the range from – 200 ppm/°C to + 200 ppm/°C. Thus they can be designed to compensate for temperature dependencies of other components in the circuit, and, for example, oscillators can be built with minimum frequency drift with temperature. The dissipation factor is very small on the order of .02% to .1% and remains low into the high MHz range.

---

*DuPont Registered Trademark.

DIELECTRIC PERFORMANCE COMPARISONS

TEMPERATURE °C

	−55	−30	0	25	45	65	85	105	125
DIELECTRIC CONSTANT @ 1 KHz (Typical)									
Mylar*	—	—	—	3.1	—	—	—	—	—
Polystyrene	—	—	—	2.5	—	—	—	—	—
Polycarbonate	—	—	—	2.8	—	—	—	—	—
Polypropylene	—	—	—	2.4	—	—	—	—	—
Aluminum Oxide	—	—	—	7.0	—	—	—	—	—
CAPACITANCE CHANGE % @ 1 KHz (Typical)									
Mylar*	−5	−2	−1	0	1	1.5	2	7	15
Polystyrene	1	.7	.3	0	− .3	− .6	− .9	—	—
Polycarbonate	−1.0	− .30	0	0	0	0	0	− .13	.20
Polypropylene	+1.4	+1.2	+ .6	0	− .4	−1.6	−2.5	−3.5	—
IR MEGOHM-MICROFARADS (WVDC 2 min. Typical)									
Mylar*	>10^5	>10^5	>10^5	10^5	4×10^4	7×10^3	10^3	4×10^2	7×10
Polystyrene	>10^6	>10^6	>10^6	10^6	7×10^5	5×10^5	4×10^4	0	0
Polycarbonate	>10^6	>10^6	>10^6	4×10^5	3×10^5	2×10^5	10^5	5×10^4	10^3
Polypropylene	>10^5	>10^5	>10^6	10^5	4×10^4	7×10^3	8×10^2	10^2	—
DISSIPATION FACTOR % @ 1 KHz (Typical)									
Mylar* @ .1 μf	1.0	1.1	.70	.32	.16	.10	.15	.52	.96
Polystyrene @ 2000 ρf	.07	.08	.04	.03	.05	.07	.08	∞	∞
Polycarbonate @ .1 μf	.45	.30	.20	.07	.08	.08	.10	.15	.16
Polypropylene @ 2000 ρf	.06	.06	.06	.06	.06	.06	.07	.06	—
DIELECTRIC ABSORPTION % (Typical)									
Mylar*	—	—	—	.20	—	—	—	—	—
Polystyrene	—	—	—	.02	—	—	—	—	—
Polycarbonate	—	—	—	.08	—	—	—	—	—
DC WORKING VOLTAGE VOLTS/GA# (Typical)									
Mylar*	8	8	8	8	8	8	8	6	4
Polystyrene	3	3	3	3	3	3	3	0	0
Polycarbonate	5	5	5	5	5	5	5	5	5
Polypropylene	8	8	8	8	8	8	8	6	−

FREQUENCY Hz @ 25°C

		60	120	10^3	10^5	10^6
DISSIPATION FACTOR % @ 25°C						
Mylar*	@ .1 μf	.20	.23	.32	1.1	10
Polystyrene	@ 500 ρf	.01	.01	.01	.01	.04
Polycarbonate	@ .1 μf	.05	.06	.07	.16	4
Polypropylene	@ 500 ρf	.01	.01	.01	.01	.13
Q FACTOR @ 25°C						
Mylar*	@ .1 μf	500	435	313	91	10
Polystyrene	@ 500 ρf	10000	10000	10000	10000	2500
Polycarbonate	@ .1 μf	2000	1670	1430	625	25
Polypropylene	@ 500 ρf	10000	10000	10000	10000	770

*DuPont Registered Tradem

Table 2-1 Film Capacitors

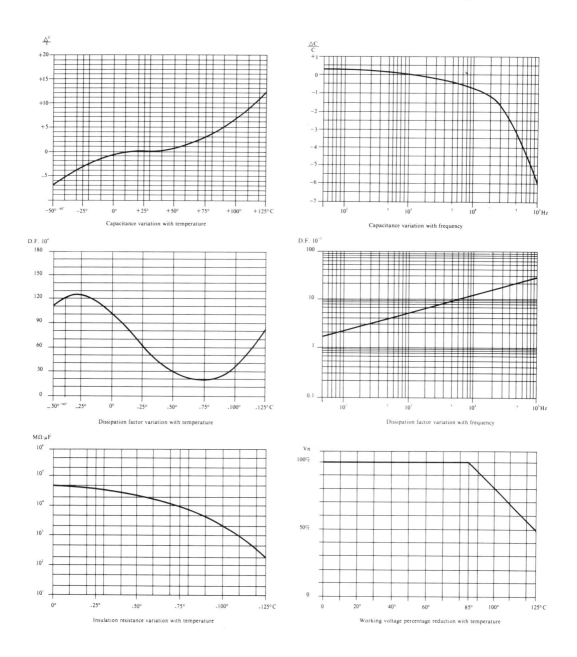

Figure 2-8. Typical performance characteristics of metalized polyester capacitors.

Two kinds of mica capacitors are available. The *metal foil* kind which consists of alternate layers of mica and metal foil, and the *silvered mica* kind which consists of mica sheets with silver deposited directly on the mica. The latter type results in lower temperature coefficients and also in greater stability during operation.

Electrolytic Capacitors

The outstanding feature of the electrolytic capacitor is its high capacitance to volume ratio. Common electrolytic capacitors are available in the range from $1 \mu F$ to 0.1 F with voltage ratings up to 500 volts. The higher capacitance values are available only at low voltage rating. For high voltage, high capacitance needs, several units must be paralleled. Shortcomings of electrolytic capacitors are their wide tolerance range from −20 to +100%, their high Equivalent Series Resistance ESR, and high d.c. leakage currents.

Electrolytic capacitors are listed as aluminum or tantalum types, dependent upon the electrode material. Both come as polar or non-polar capacitors. The polar types are by far the more common. The voltage can be applied only with the specified polarity. If the polarity is reversed, the leakage current is excessive, causing overheating and possibly the explosion of the capacitor. The non-polar electrolytic capacitor can be connected to a voltage with either polarity. The non-polar capacitor, however, has only half the capacitance of a polar capacitor of the same volume. The dielectric in an electrolytic capacitor is an oxide grown on the electrode metal by anodization. In the polar capacitor only one electrode is anodized, while both electrodes are anodized in the non-polar capacitor. If the wrong polarity has been connected momentarily to a polar capacitor, or if the capacitor has been stored unused for several months, it may be necessary to form the dielectric layer by the application of the dc working voltage with the proper polarity until the leakage current reduces to the nominal value.

Relatively small electrolytic capacitors are made in tubular form with one wire extruding from either end with the positive terminal clearly marked. Larger valued capacitors can be contained in one can with or without one common electrode.

A miniature type of electrolytic capacitors is the solid tantalum type molded into an epoxy. Capacitance values range from a fraction of a microfarad to several hundred microfarads.

Trimmer Capacitors

Trimming capacitors are variable capacitors used for a one-time screwdriver adjustment to an exact desired value. The adjustment is normally done experimentally to peak a resonant circuit, for example, or to adjust a band-pass filter, or a time-constant. Commonly used are small screw-mounted ceramic capacitors on the order of $1/2'' \times 3/4''$ in size Capacitance values range from a picofarad to several hundred picofarads. The adjustable range of any one unit covers a range of 4:1 to 10:1. Examples are 1 to 4 pF, 8 to 50 pF and 10 to 100 pF. Voltages are in the range from a few hundred to a few thousand volts with dissipation factors on the order of .2% at 1 MHz.

Feed-Through Capacitors

Feed-through capacitors are used in high frequency circuits where it is desired to bypass to ground a wire passing through a shield-box. The capacitor provides terminals on both sides of the wall and a capacitance to the wall. Two types are available, a soldered type and a thread-and-nut mounted type. Capacitance values range from 100 to 5000 pF.

Air-Variable Capacitors

These are used where a tunable resonance circuit is required such as for the front-end of a receiver. It consists of interleaved sets of metal plates, one set stationary and the other set movable by means of a rotating shaft. The dielectric is air. By rotating the shaft, the capacitance can be varied between a few picofarads and a few hundred picofarads.

2.3.3   Pitfalls to Avoid

Often a high valued capacitor is used to bypass relatively low frequencies. Such a capacitor, however, may not bypass high frequencies (which one would expect to be bypassed, based on circuit RC time constant consideration for an ideal capacitor) because the ESR of electrolytic capacitors becomes excessive at high frequencies. The remedy in such a case is to parallel the capacitor with a small value (0.1 μF) high frequency capacitor.

2.4   INDUCTORS

When designing an electronic circuit or system with I.C.'s one tries to avoid the use of inductors because of their bulkiness. Tuned circuits can be designed using certain R-C bridge networks, but at times inductors cannot be avoided. At very high frequencies, inductors are not a cause of much concern because the few microhenries needed can be obtained by bending a wire into a few-turns, self-supporting air inductor. Important factors to consider are the inductance, the resistance, the resulting Q value and the maximum d-c current rating.

The d-c current rating is meaningful only for iron core inductors and corresponds to the current value which will drive the iron core into saturation, resulting in reduced inductance. The quality factor Q is the ratio of the coil reactance X_L, which depends on frequency, to the d-c resistance R_{dc}:

$$Q = \frac{X_L}{R_{dc}} = \frac{\omega L}{R_{dc}}$$

where ω is the angular frequency and L is the inductance.

The effective a-c resistance can also be a function of frequency as a result of eddy current and hysteresis losses in the core. This results in the broadening of the bandwidth of a tuned circuit and is sometimes used deliberately for this purpose.

2.4.1 Air Core Inductors

At very high frequencies, in the range of 100 kHz to 1 GHz, the needed inductances are usually on the order of a few tenths of a microhenry to several hundred microhenries. Inductors with these values can be obtained as *air coils*. Such coils are either self-supported or wound on insulating cylindrical forms. At the higher frequencies where very low inductance values are needed, the inductors can be handmade by forming a few loops of wire. We shall illustrate this with an example.

The inductance of an air coil is given by:

$$L = \frac{r^2 n^2}{9r + 10l}$$

where

    L is the inductance, in microhenries;
    r is the radius of the coil, in inches;
    l is the length of the coil, in inches;
    n is the number of turns of the coil.

Suppose we want to construct a coil of 2 μH inductance. The wire diameter is determined by the current that it might carry and by the required strength and rigidity (if it is self-supporting). Suppose that a wire diameter of 0.005 in. is suitable. The length of the coil can then be expressed as l = 0.005n. We choose a coil radius of 0.25 in. The equation for the inductance then becomes:

$$2 = \frac{0.25^2 \times n^2}{9 \times 0.25 + 10 \times 0.005 \times n} = \frac{0.125n^2}{2.25 + 0.05n}$$

Solving for n we obtain only one physically realizable solution, n = 6.8. (The other solution for n is negative.) Thus, a 7-turn coil is the solution. The coil can be stretched slightly lengthwise to obtain just the right value. In practice, the needed value would be obtained by expanding or compressing the coil slightly to obtain the correct functioning of the circuit in which the coil is being used, rather than by adjusting the coil independently.

2.4.2 Iron Core Inductors

Iron Core Inductors are available commercially in a variety of physical sizes and electric characteristics. Most desirable for designs with integrated circuits are small size, so-called *miniature*, inductors. Typical case sizes are less than 1/2-inch diameter and 1/4-inch height. Inductance values at zero dc current range typically from 0.1 to 120 mH, and dc resistance values range from a few ohms for the lower valued inductors to several hundreds and thousands ohms for the higher inductance coils. The maximum permissible dc current may be as low as 0.2 mA or as high as 90 mA. Q values of about 40 at 10 kHz are available. There are also available high-Q inductors at low frequency, but they are larger in physical size. For example, a 2 Henry inductor with a case size of 1.3 in.-diameter and 1.6 in.-high may have a Q value of 40 at 60 Hz.

2.5   TRANSFORMERS

Transformers are used in power supplies primarily to obtain the needed voltage levels, in IF and RF stages primarily for circuit tuning, in video and audio stages primarily for impedance matching and in pulse circuits for voltage pulses transmission. In addition to providing these primary functions, transformers can also provide electric isolation, polarity inversion, impedance level change for noise reduction and for other functions. It should be noted, however, that in modern design with I.C.'s, the use of transformers is rather scarce.

2.5.1 Power Supply Transformers

When deciding on the use of a power supply for a particular electronic circuit, the designer can choose between a ready-made packaged power supply, and I.C. voltage regulator, and the design of a power supply using individual components. The first choice results in the most expensive power supply, but it is ready for use with input terminals for connection to

the 115 volt line and with output terminals for the desired regulated d-c
voltage. A selection of regulated voltage levels and polarities, as well as
current levels from a few milliamperes to several amperes are available. Off-
the-shelf I.C. voltage regulators are available up to a power output of 5W.
The packages must be mounted on a heat sink and the designer must provide
the unregulated d-c voltage to the voltage regulator. This involves the design
of a circuit which contains a power transformer, rectifiers and capacitors.
Finally, a similar circuit must be designed if the designer assumes the
responsibility for the design of the complete regulated power supply.

The selection of a power transformer is based on voltage ratings and
ratios, frequency, power rating, and winding resistance. The primary side
voltage must correspond to the line voltage, which usually is 115V, and the
specified frequency must correspond to the line frequency, which usually is
60 Hz. A frequency of 400 Hz is common in some military applications.
Transformers are available with a frequency range specification from 50-
5000 Hz, but usually the frequency specification is 50-60 Hz or 400 Hz. The
voltage that can be supported by a transformer winding is given by the
relation:

$$V = 4.44f \, N \, \phi \times 10^{-8}$$

where V is the rms voltage (in volts) of a sinusiodal voltage, f is the frequency
(in Hertz), N is the number of turns, and ϕ is the magnetic flux (in
Maxwells). For square wave applied voltages, the number 4.44 must be
replaced with the number 4 and V is the peak voltage. The designer selecting
a transformer need not concern himself with this question, but it is
convenient to use it to explain some simple, but important, facts. If the rated
voltage is applied, but the frequency is too low, the flux ϕ will increase above
the nominal value causing saturation of the magnetic core and overheating
of the transformer. Also, the impedances of the coils will change with
frequency.

The equation also shows that for a given value of a-c flux, at a given
frequency, the voltage across any given coil is proportional to the number of
turns; the ratio between the voltages of any two coils is therefore
proportional to the turn ratio. The secondary voltage of a power
transformer must be chosen according to the desired value of the d.c.
regulated voltage. The peak value of the secondary voltage must be higher
than the desired d.c. regulated voltage. In some power transformers there
are two secondary windings which can be connected either in series or in
parallel, providing two possible values for the secondary voltage. Standard
voltages are 5, 10, 12, 14, 20, 24, 27, 30, 33, 36 and 40 rms volts. Thus,
transformers are available for any desired d.c. regulated voltage level
suitable for I.C.'s and transistors. Higher voltages are also available for

transistors requiring hundreds of volts. These are usually listed under power transformers for vacuum tubes. The low level voltage transformers are small in size having linear dimensions on the order of 1″ to 1 1/2″, and often have lug connections suitable for printed circuit mounting.

The power rating is given either as a VA (voltage × current) rating or in terms of output current, which when multiplied by the output voltage gives the power. The value of the winding resistance is important since it affects the regulation. To obtain the voltage output under full load, the product of the resistance times the full load secondary current must be subtracted from the nominal output voltage. The resistance must be taken as the sum of the secondary winding resistance plus the product of the secondary to primary turn ratio times the primary resistance.

Since power transformers are used in conjunction with rectifiers, some manufacturers specify the output voltage in terms of the d.c. voltage when the transofrmer feeds either into a half-wave rectifier or into a full-wave rectifier. Thus, for example, the secondary voltage of a transformer may be rated at 9 volts d.c. for a 1/2-wave rectifier, and 18 volts d.c. for a full-wave or bridge rectifier. These specifications apply to the circuits shown in Figures 2-9 (a) and 2-9(b), respectively. Higher voltages may be obtained through the use of capacitor input filters. In this case, however, the rated d.c. current must be reduced by a factor of 2. If a voltage doubler is used, the current must be reduced by a factor of 4.

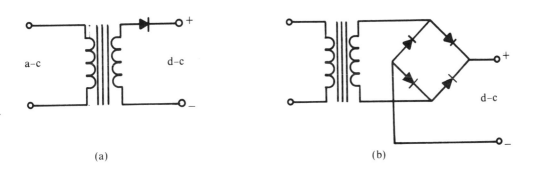

(a) (b)

Figure 2-9. Voltage rectifier circuits for power supplies.

(a) half-wave rectifier

(b) full-wave rectifier

2.5.2 Audio Transformers

The important characteristics of audio transformers are the power rating, frequency response and impedance levels. The power rating must be compatible with the load, such as a speaker. Similarly, the impedance levels must be chosen to match the output impedance of the amplification stage to the input impedance of the load for maximum power transfer. If, for example, the output impedance of the amplification stage is 1 kohm and the speaker impedance is 4 ohms, then an audio transformer with a 1 kohm primary impedance and 4 ohm secondary impedance is suitable. The frequency response of an audio transformer is defined as the frequency range over which the voltage attenuation is less than 3dB when terminated with the specified impedance levels. A typical frequency range is 50 to 20,000 Hz. The upper cut-off frequency is limited by the series inductance and shunt capacitance of the windings and the lower cut-off frequency is limited by the need for increased magnetizing current at lower frequencies resulting from the decrease in primary reactance. Eventually, the voltage drop across the source and primary winding resistance becomes excessive causing a significant drop in the secondary voltage.

Audio transformers can also be used to couple audio amplification stages in which case the impedance levels may be, for example, 20 kohms and 800 ohms. For use with I.C.'s, audio transformers come encapsulated in cylindrical and rectangular shapes with all the wires coming out from one side or from two sides for use with printed circuit boards.

2.5.3 Video Transformers

Video transformers are wide-band devices, that is, their frequency response extends into the megahertz range. Transformers are available with a bandwidth from 20 Hz to several megahertz. The response of the transformer in a system is intimately related to the source and load impedances, and to the wiring, as the frequency dependent behavior is very sensitive to wiring inductances and capacitances at these high frequencies.

2.5.4 I.F. and R.F. Transformers

I.F. and R.F. transformers are used to couple and tune circuits in the hundreds of kilo-ohms and in the megaohm range. Unlike audio and power transformers, these transformers are not wound on a closed magnetic core; the primary and secondary windings are wound concentrically on a bobbin which contains a movable iron or copper slug inside. The mutual inductance between the coils can be adjusted by lowering or raising the slug; lowering an iron slug into the coils increases the inductance and lowering a copper slug decreases the inductance, since the permeability of copper is smaller than the permeability of air. Each winding has a capacitor connected across it. Adjustment of the metal core by means of a screwdriver adjusts the center (peak) frequency. Instead of a variable inductance, the transformer can have screwdriver capacitor adjustments. The primary and secondary circuits can be tuned to the same frequency for sharp and narrow tuning, or to different frequencies for broader band-pass tuning. Similarly, the degree of mutual coupling, by means of the metal slug adjustment, changes the frequency response from a narrow band single peak to a wider band, double peak characteristic. The fine tuning is done under active operating conditions, that is, when the system is *on*. The important specifications of an I.F. or R.F. transformer are the center frequency, bandwidth range and primary and secondary impedance levels.

2.5.5 Pulse Transformer

Pulse transformers are used to couple voltage pulses to change voltage levels or to isolate electrically one part of a circuit from another. The important specifications of a pulse transformer are, in addition to the voltage and power levels, the *rise time, fall time, overshoot, undershoot,* and *sag* as shown in Figure 2-10.

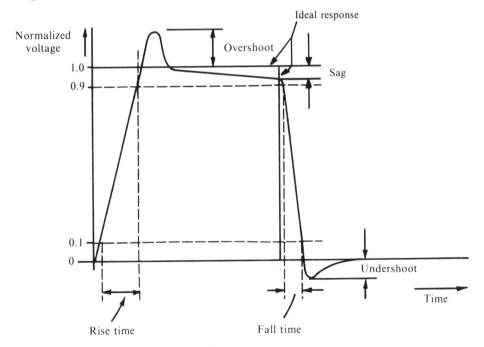

Figure 2-10. Pulse transformer response curve.

Chapter 3

Designing Low Noise Circuits

3.1 In this chapter we discuss the electronic noise associated with the elements of our circuits. This noise is generated by the motions of electrons in passive and active devices and it sets a lower limit on the a.c. signal level which we can accurately resolve. Because the electron is a discrete particle, all d.c. currents and voltages are the average effect of many small charge movements. At any given instant, the actual current or voltage may differ from the average value. Clearly, since the measured signals at various instants may differ, we must associate an a.c. component with any d.c. current or voltage. This a.c. signal we call noise.

The effects of noise are of interest in low level preamplifier stages where the output of a transducer, e.g., a phonocartridge, photodiode, microphone, etc. is amplified for subsequent signal processing. In such systems the noise represents an unwanted signal which can "mask" or obscure the desired information. Since the origins of noise are fundamentally linked to the nature of the electron we can never completely eliminate noise signals, but through proper design we can minimize their effect. This is the goal of the present chapter.

Excluded from this chapter are the methods for reducing the effects of spurious signal pickup which results when ineffective shielding, poor wiring layout, or ground loops allow stray fields or signals to affect a sensitive circuit. These problems are treated in Chapter 11.

For many years transistor circuits continued to be the best choice for low noise applications, but recent processing advancements have produced a generation of Low Noise I.C.'s which have few discrete rivals, especially in cost-effective designs. Furthermore, the small I.C. package size and reduced component count result in a circuit which is more compact and therefore more easily isolated from spurious signals. These factors combine to make a properly selected I.C. the best choice for most low noise applications.

103

3.2   NOISE: PROPERTIES AND MEASUREMENT

Before discussing particular noise sources we must establish a way to measure, and thus characterize, a noise source with a number or a graph. If we look at the output of a high gain preamplifier with an oscilloscope we see a jagged or fuzzy line centered about some d.c. level. If the amplifier has no input signal, we call this trace *noise*. The output of another amplifier may be more or less fuzzy. Although the jagged oscilloscope trace forms our common notion of noise, this *time-domain* representation does not lend itself to easy measurements or useful characterization. The answer lies in displaying the noise in the *frequency-domain*. This amounts to looking at the noise voltage through a series of narrow band filters and determining how much energy the noise source has in each frequency band. The circuit of Figure 3-1 performs this function and is the basic noise measuring circuit. Although an amplifier is shown in this example, any device, active or passive, that we suspect of being noisy can be measured in this manner. Circuit operation is as follows. A sweep generator slowly moves the center frequency of a 1 Hz bandwidth filter. At each frequency f_x the output of the filter contains only the components of the input noise signal which fall in the 1 Hz interval about f_x. This a.c. signal is converted to a d.c. level which is proportional to the power of the original a.c. signal and is then applied to the vertical input of an x-y recorder whose x-axis corresponds to the center frequency f_x. Thus, a graph is produced with V^2/Hz as the y-axis and the frequency is the x-axis.

For reference, consider applying a pure sine wave at 1 kHz to this system. The filter would begin sweeping, but until 1 kHz was reached the output would be zero. At 1 kHz the filter output is the same as the input and the power measurement circuit would compute the power in the wave and position the recorder to this value. From 1 kHz on the output is again zero. The resultant display is shown superimposed in Figure 3-1. A comparison of the single frequency display and the noise signal suggests that the noise consists of a continuum of frequencies. In fact, theoretical calculations show that the truly random voltage pattern which we associate with noise must have a frequency distribution which is continuous. Notice that the unit of noise on this graph is volt $^2/Hz$. Thus, if we want to calculate the noise in a given bandwidth, we simply find the area under the curve between the two frequency limits. The resulting unit will be $(Volt)^2$ and is a measure of the relative total noise power in that bandwidth. The noise voltage can be found by taking the square root of the noise power, i.e.:

$$\sqrt{\frac{V^2}{Hz}} \;=\; \frac{V}{\sqrt{Hz}} \quad \text{and is an r.m.s. value.}$$

The frequency-domain noise representation has proved most useful

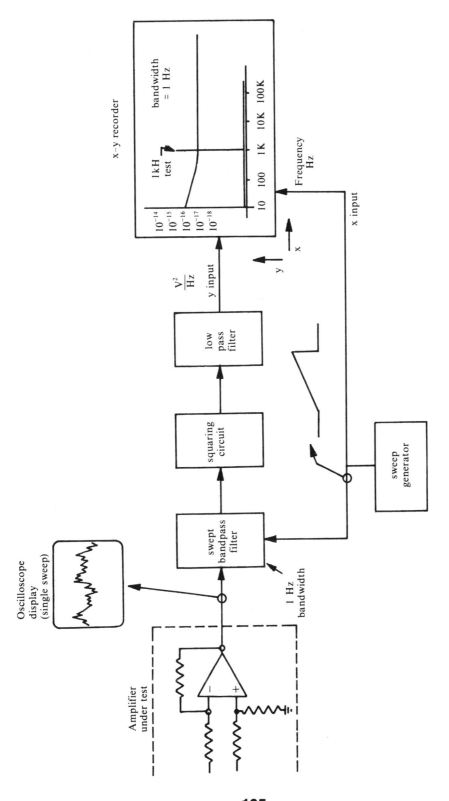

Figure 3-1. Noise measuring circuit.

in the majority of noise problems. However, the time-domain representation is the real time picture and in certain types of low noise problems, e.g., pulse detectors, we need to know more about the character of the noise than the frequency-domain representation can tell us. Some of these special problems will be treated later in the chapter.

The following conventions will be used:

1) *Noise Densities* are expressed in *lower case letters:*
Examples:

v_n = rms noise voltage density in a one-Hertz bandwidth

$$\text{Units} - \frac{\text{Volts (rms)}}{\sqrt{\text{Hz}}}$$

$\overline{v}_n^2$ = (read "v bar subscript n squared") = relative[1] mean square noise power density in a one-Hertz bandwidth.

$$\text{Units} - \frac{\text{Volts}^2}{\text{Hz}}$$

2) *Total Noise* in a given bandwidth [2] $\overline{B}$ (read "B bar") is expressed in *upper case letters*
Examples:

V_n = total rms noise in $\overline{B}$

$$\text{Units} - \text{Volt (rms)}$$

$\overline{V}_n^2$ (read "V bar subscript n squared") = total relative[1] mean square noise power in $\overline{B}$

$$\text{Units} - \text{Volt}^2$$

The *Noise Density* and *Total Noise* are very simply related if the noise density is not a function of frequency. Specifically:

$$Vn = v_n \times \sqrt{\overline{B}}$$

and

$$\overline{V}_n^2 = \overline{v}_n^2 \times \overline{B}$$

---

[1]The term "relative" is used since the actual power is not expressed. To determine the power, we need to know the resistance across which the voltage is measured.
[2]The bandwidth is defined as: $B = (f_H - f_L)$ where f_H is the upper frequency limit and f_L is the lower frequency limit.

Since many noise sources are independent of frequency, so-called *white noose,* these equations are quite useful.

For *Noise Current* we have:

i_n = rms noise current density in a one Hertz bandwidth

$$\text{Units} - \frac{\text{Amps}}{\sqrt{\text{Hz}}}$$

i_n^{-2} = relative mean square noise power in a one Hertz bandwidth

$$\text{Units} - \frac{\text{Amp}^2}{\text{Hz}}$$

I_n = total rms noise in bandwidth $\overline{B}$

Units—Amps (rms)

$\overline{I}_n^2$ = total relative mean square noise power in $\overline{B}$

$$\text{Units} - \text{amp}^2$$

In noise calculations involving multiple sources remember:
1) Noise powers add algebraically. Thus:

$$\overline{V}_{total}^2 = \overline{V}_{n1}^2 + \overline{V}_{n2}^2 \ldots \tag{3-1}$$

and

2) RMS noise voltages and currents add vectorially. Thus:

$$V_{total} = \sqrt{(v_{n1})^2 + (v_{n2})^2 + \ldots} \tag{3-2}$$

The use of these equations is illustrated in numerous examples throughout this chapter.

3.3   PASSIVE NOISE SOURCES

There are only three possible noise sources in electronic circuits: the resistor, the direct current and the active elements. Ideal capacitors and inductors produce no noise, but they can affect overall noise performance by attenuating the desired signal and thus changing the signal to noise ratio. However some real capacitors (see Chapter 2), notably electrolytics, have some undesirable effects and, of course, inductors are prone to pick up stray fields. The ideal transformer is also noiseless, but due to its impedance transforming ability it is useful in some low noise designs (see Section 3.3.8).

Emphasis is placed on understanding the noise characteristics of resistors and d.c. currrents since the noises associated with most active elements are generated by resistances and currents within the active device (e.g. transistor). Let us examine each noise source in detail.

3.3.1 The Resistor

In the resistor, thermally excited electron currents flow in the resistance element and, though random in nature, they sum statistically to produce a noise voltage across the resistor. This effect is variously called *Thermal, Johnson,* or *Nyquist* noise. The noise power density $\overline{v}_n^2$ is described by:

$$\overline{v}_n^2 = 4kTR \left(\frac{Volts^2}{cycle}\right) \tag{3-3}$$

where:

K = Boltzmann's constant = $1.33 \times 10^{-23joules}$
T = Temperature in $°K$
$4KT = 1.6 \times 10^{-20}$ Joules
R = value of the resistor (Ohms)

This equation states that the square of the noise voltage in a bandwidth of 1 Hz is a function of temperature and resistance. To calculate the total noise power $\overline{V}_n^2$ in a given bandwidth $\overline{B}$, we multiply the value $\overline{V}_n^2$ by $\overline{B}$. Thus:

$$\overline{V}_n^2 = 4kTR\overline{B}. \tag{3-4}$$

To get the root mean square voltage we take the square root,

$$V_n = \sqrt{4kTR\overline{B}} \tag{3-5}$$

A frequency-domain plot of the resistor noise would look like Figure 3-2. Here we have chosen room temperature and $1M\Omega$ resistance.

Since the noise voltage in a given bandwidth is independent of frequency, the graph is a flat line. For example, to find the total rms noise voltage in a 5 kHz bandwidth produced across a 1 Meg resistor due to thermal effects, at $T=300°K$, we have by Equation 3-5,

$$V_n = \sqrt{4kTR\overline{B}} = \left[(1.6\times10^{-20})(1\times10^6)(5000)\right]^{1/2} = 8.9\mu \text{ Volts rms}$$

Notice only the bandwidth and not the frequency itself appears in this equation. Since Equation 3-3 states that the noise power is proportional to bandwidth, the rms noise voltage is proportional to the square root of the bandwidth. Thus, the introduction of the odd looking character $\sqrt{Hz}$ into noise descriptions. Noise which has the flat frequency characteristic of Figure 3-2 is called *white noise* by analogy to white light which contains all light frequencies in equal amounts.

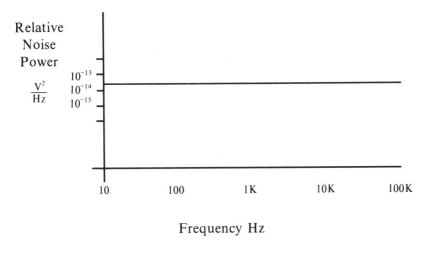

Figure 3-2. Noise power density of a resistor. R = 1M.

Using Equation 3-5 we can generate the family of curves of Figure 3-3. Using this graph, we can find the total rms noise voltage for any resistor at any bandwidth for T=300°K (room temperature).

Example: Find V_{rms} for R=1 kΩ and B=100 kHz

Step 1: locate 100 kHz on the x-axis.
Step 2: Move vertically up to the 1 kΩ line.
Step 3: move horizontally to the y-axis and read the total rms noise voltage, 1.26μ Volts.

One is tempted to ask, if a resistor has a noise voltage at its output terminals, why can't we hook it up to a load and get some work out of it? In fact, we can—but there is a catch. Let us rewrite Equation 3-4 as:

$$\frac{\overline{V_n^2}}{R} = 4kT\overline{B} \qquad (3\text{-}6)$$

On the left side of the equal sign is the familiar expression for the power dissipated in a resistor, and since the right side is dependent on temperature and bandwidth only, this equation states that *all resistors* at the

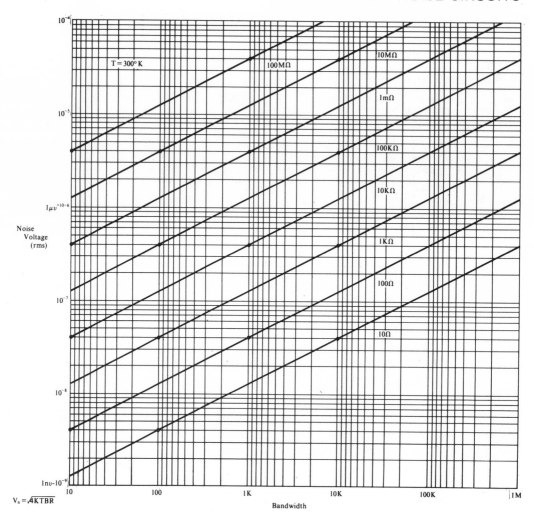

Figure 3-3. V_n verses B for various values of R.

same temperature are generating the *same noise power* assuming equal bandwidths. Now, if we hook up two resistors as in Figure 3-4 we see that if the resistors are at equal temperatures, then each resistor supplies identical power to the other resistor and no overall energy transfer takes place. However, if one resistor is at a higher temperature than the other, then an actual transfer of energy would take place and the situation is reduced to a thermodynamic problem where energy transfer occurs only where temperature differentials exist.

This example serves to emphasize the thermodynamic origin of the noise itself and implies that we can characterize the noise of a given resistor R by specifying the noise voltage as in Equation 3-3 *or* just as good a

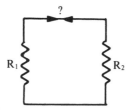

Figure 3-4. Can one resistor supply noise power to another?

description can be given by specifying the noise current. Since V=IR, Equation 3-3 can be written as:

$$\overline{I}_n^2 = \frac{4kTB}{R} \ \text{Amps}^2 \tag{3-7}$$

$$I_n = \sqrt{\frac{4kT\overline{B}}{R}} \tag{3-8}$$

Depending on the particular noise problem the noise voltage density v_n or the noise current density i_n proves most useful. As an aid to calculating this current, we can generate the set of curves of Figure 3-5. Figures 3-3 and 3-5 are used in a similar manner.

Where multiple resistors appear in series or parallel arrangements, we first reduce the network to an equivalent resistor and calculate the noise voltage or current using the equivalent resistor value and Equation 3-4 or Equation 3-7.

As an example, calculate the rms noise voltage and noise current over the audio bandwidth (10Hz-20KHz) for the network of Figure 3-6 at room temperature.

We easily find the equivalent resistor (2M) and from the graphs of Figures 3-3 and 3-5 or equations 3-4 or 3-7 we find:

$$V_n = .17 \ \text{Volts}$$

$$\text{For } T=300°K; \ \overline{B}=20kHz - 10Hz \approx 20kHz$$

$$I_n = 120p \ \text{Amps}$$

Remember, the noise voltage must be read with a high impedance meter. In this example R_{in} (meter input resistance) must be greater than 2M. The noise current must be read with a low impedance meter; R_{in} much less than 2M, the meter circuit being that of Figure 3-1.

The resistor noise description we have given so far is valid for ideal resistors. Real resistors, the ones we manufacture, have additional noise

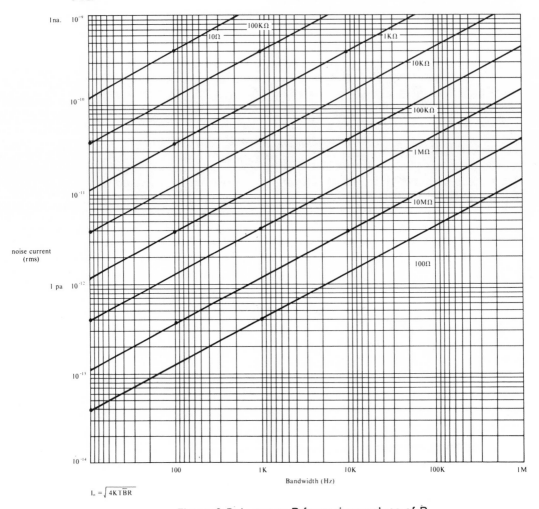

$$I_n = \sqrt{4KTBR}$$

Figure 3-5. I_n versus B for various values of R.

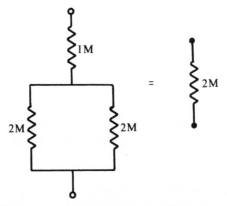

Figure 3-6. Equivalent Noise Resistance.

components which are not explained by the simple thermodynamic description we have assumed. However, certain types (see Chapter 2) approach the ideal performance so closely that no corrections to Equations 3-4 and 3-7 are needed. Unfortunately, the common carbon composition resistor tends to be noisier than the ideal resistor and should be avoided in critical low noise applications.

3.3.2   Shot Noise

The d.c. current is the second important noise source. It is called *shot* or *Schottky* noise. Again, the discrete nature of the electron is the cause.

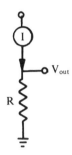

Figure 3-7. Shot noise.

Figure 3-7 shows a current source sending current into a noiseless resistor R (This is a perfect resistor, not an ideal one!). The output impedance of the current source must be much greater than R. Because the current is composed of discrete electrons, the voltage across the resistor is actually the sum of the many small pulses associated with the arrival of individual electrons. Theory and measurement confirm that this process results in white noise, i.e., the noise energy in a given bandwidth is independent of frequency. The noiseless resistor R. helps to illustrate a point, but since (V=IR) we can refer the output noise voltage back to a noise current associated with the d.c. current. In practice, shot noise should be imagined as a noise current. The square of the noise current in a one-cycle bandwidth associated with the direct current I_{dc} is given by:

$$\overline{i_n}^2 = 2qI_{dc} \tag{3-9}$$

where, q=electronic charge=$1.6+10^{-19}$Coulomb.

Again we start with $\overline{I_n}^2$ because this is the quantity we measure with a circuit similar to that shown in Figure 3-1. Then, for a given bandwidth $\overline{B}$ Equation 3-9 becomes:

$$\overline{I}^2_n = 2q\ I_{dc}\ \overline{B} \text{ and for the rms value}$$

we have

$$I_n = \sqrt{2q\ I_{dc}\ \overline{B}} \tag{3-10}$$

A graph of this equation appears in Figure 3-8.

    The following example illustrates the use of Equation 3-10. Figure 3-9 is a simple optical communication receiver. The photodiode is reverse-biased and with no light incident on the diode, the leakage current I_D (dark current) flows into resistor R_1. Ignoring for the moment the noise associated

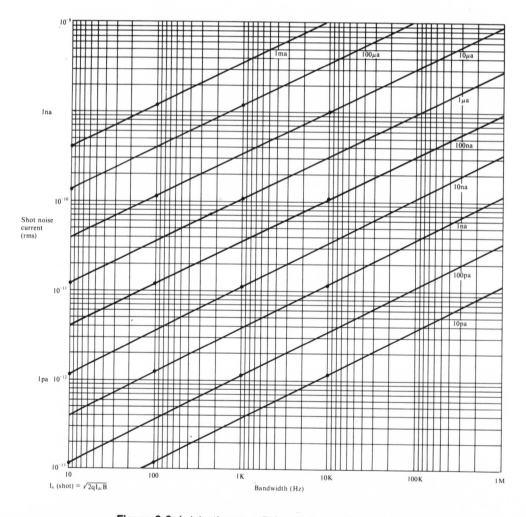

Figure 3-8. I_n (shot) versus B for various values of I_{dc}.

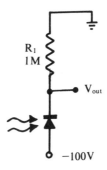

Figure 3-9. Infrared detector.

with R_1, calculate the rms noise voltage which appears at the output V_{out} due to the shot noise associated with the leakage current I_D. Assume $\overline{B}=10\,\mathrm{kHz}$ and $I_D=3\times10^{-9}$

First we calculate I_n, from Equation 3-10,

$$I_n = \sqrt{2q\ I_{dc}\ \overline{B}}$$

$$I_n = \left[(2)(1.6\times10^{-19})(3\times10^{-9})(10^4)\right]^{1/2} = 3.1\ \mathrm{pA.\ rms}$$

and from V=IR, if $I=3.1\times10^{-12}$A and R=1M, then at the output, $\overline{v}_{rms}=3.1\times10^{-12}\times 1\times10^6=3.1\mu$ volts rms.

The thermal noise of R_1 which we have ignored is in fact a very critical noise source and this circuit is examined later.

3.3.3   Resistors and D.C. Currents

Excluding active devices, we can characterize the noise performance of a circuit by calculating the shot noise and thermal noise components and summing them. In a particular low noise design we may wish to know which noise source, shot or thermal, is the limiting factor. A surprisingly simple solution exists.

Consider the circuit of Figure 3-7. The output noise voltage consists of two components:

1) the shot noise, which is given by:

$$(V_n)_{shot} = I_n\ R = \sqrt{2q\ I_{dc}\overline{B}} \times R$$

and

2) the thermal noise which is given by:

$$(V_n)_{thermal} = \sqrt{4kTRB}$$

Let us calculate the conditions necessary for the thermal and shot noise contributions to be equal. Let:

$$\overline{(V_{rms})} \text{ shot} = \overline{(V_{rms})} \text{ thermal}$$

Then,

$$\sqrt{2q \, I_{dc} \, \overline{B}} \times R = \sqrt{4kTR\overline{B}}$$

squaring both sides, canceling B, and rearranging terms we have:

$$I_{dc} \times R = \frac{4kT}{2q}$$

but $I_{dc} \times R$ is just the d.c. voltage across the resistor R; so we have:

$$V_{dc} = \frac{4kT}{2q} = .05 \text{ Volts at room temperature} \qquad (3\text{-}11)$$

This simple result states that if the voltage V_{dc} produced across *any resistor* by a d.c. current is equal to 50 mV, then the total noise voltage across that resistor is composed of equal amounts of shot noise and thermal noise. This is an extremely general and useful result. Recall that we must live with thermal noise (unless we lower the temperature) but presumably we have some control over the d.c. current levels in our circuits. Therefore, in the design of low noise circuits we should minimize d.c. currents to reduce the shot noise below the thermal noise. Equation 3-11 provides a simple check of our shot noise levels; if the voltage across a resistor is greater than 50mV, then shot noise dominates and if the voltage is less than 50mV, thermal noise dominates (for T=300°K). To obtain this result, we assumed that I_{dc} was supplied by a current source. If this is not the case, some simplifying techniques can be applied (see Example 1, below).

The relatively low voltage $(.05V_{dc})$ across a resistor at which equal noise contributions from shot and thermal effects occur, emphasizes the importance of considering shot noise in all noise calculations. We commonly separate the a.c. and d.c. portions of our circuits, but if shot noise is an important source, this separation may not be possible. The d.c. current creates a noise spectrum independent of frequency and although we may block the d.c. current with a capacitor or a filter, the a.c. noise caused by that current is not blocked (see Example 2).

Examples: As we will see, the noise associated with active devices can, for the most part, be explained in terms of resistances and currents in the device itself. The following examples illustrate the application of the

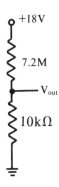

Figure 3-10. Simple bias network.

results presented so far and serve to summarize these results prior to treating active devices. All examples assume T=300°K.

Example 1: Find the total rms noise voltage/ $\sqrt{Hz}$ at the output of the divider of Figure 3-10. *Solution:* I_{dc}=2.5 μA and the equivalent output resistance of this network is just 10kΩ because 7.2kΩ>>10k. Thus,

$$(v_n)thermal = \sqrt{4kTR} = (1.6 \times 10^{-20} \times 10^4)^{1/2} = 12.6 \frac{nV}{\sqrt{Hz}}$$

$$(v_n)shot = (i_n)shot \times R = \sqrt{2q\ I_{dc}} \bullet R = (3.2 \times 10^{-19} \times 2.5 \times 10^{-6})^{1/2} \times 10^4 = 8.9 \frac{nV}{\sqrt{Hz}}$$

$$(v_n)\ total = \sqrt{(v_n)^2\ thermal + (v_n)^2\ shot} = \sqrt{(12.6 \frac{nV}{Hz})^2 + (8.9 \frac{nV}{Hz})^2} = 15.4 \frac{nV}{\sqrt{Hz}}$$

Since we have not assumed any particular bandwidth, we left the result in terms of nV per root Hz. If the bandwidth had been 10k Hz, the total noise would be:

$$(V_n)total = (v_n)total \times \sqrt{B} = 15.4 \frac{nv}{\sqrt{Hz}} \times \sqrt{10^4} = 1.54 \mu v\ rms$$

In this example, the shot noise contribution is less than the thermal noise component. How would we apply our 50mV criterion? We see that the d.c. voltage across the 10kΩ resistor is 25mV, but across the 7.2mΩ we have almost 18V and yet we are thermally noise limited. One simple way to view this is to consider that the 7.2mΩ resistor is a current source supplying I_{dc} to the 10kΩ resistor. More generally, however, we use the same approach used for calculating the equivalent resistance. In this example, we calculated the output impedance by looking back into the output terminals. In this case, we see 10kΩ in parallel with 7.2mΩ. We are, in effect, grounding all d.c. supplies and calculating the resistance between the terminals of interest, in this case the output and ground. Exactly the same technique governs the applications of Equation 3-11. First, calculate all node voltages, then ground all supplies, but consider the d.c. currents as still flowing and then apply the shot noise crossover criterion (Equation 3-11) to the node of interest. In our example, the output is the node of interest and the total voltage caused by $I_{dc}=2.5\mu A$ is 25mV. Thus, as we discovered, we are thermally noise limited. Note that if split supply voltages are used they should be treated as one supply with a total voltage equal to the sum of the individual supplies.

Example 2: This example illustrates our inability to separate certain a.c. and d.c. components in a low noise design. Figure 3-11 is an improved optical communications receiver. The transmitted beam is now an amplitude modulated carrier centered at 20 kHz. Our receiver has a band pass filter centered at 20 kHz with a bandwidth of 5 kHz. Assume the filter has unity gain in the passband and zero gain elsewhere.

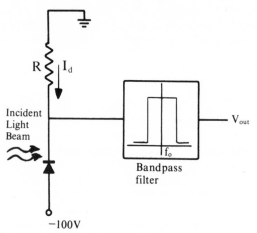

Figure 3-11. Improved IR detector with bandpass filter. Incident
light beam is amplitude-modulate at 20kHz.

First, we consider the receiver to be covered up, thus only the diode dark current I_d flows into resistor R and a noise voltage which is due to R and I_d appears at the filter input. Clearly, we desire the noise performance of this system to be limited only by the detector, i.e., photodiode noise and not by the noise associated with R. Thus, we require that the major portion of the noise voltage at the output be the result of the shot noise current, $\sqrt{2qI_d}$, flowing in resistor R. In this case, *we want to be shot noise limited* so we choose the d.c. voltage across R due to I_d to be much greater than 50mv. Let $V_{d.c.}$ =500mv. So with:

$$I_d \quad = \quad 3(10^{-9}) \text{ Amps, and}$$

$$V_{dc} \quad = \quad .5 \text{ Volts, we have}$$

$$R = \frac{V_{dc}}{I_d} = 160 \text{ Meg.}$$

This assures a photodiode noise limited system.

Remember, R is a multiplier for both the noise current *and* any signal that is incident on the receiver. Therefore, increasing R raises the voltage levels of both the signal and the noise equally at the output. The signal-to-noise ratio of the receiver itself is not changed by increasing R. However, both shot noise current *and* signal current levels increase *linearity with R* but, the thermal noise of R is increasing only as the *square root of R* since $v_n \sqrt{4kTR}$. Thus, by raising R we increase the ratio of shot noise to thermal noise until eventually the shot noise dominates. Equation 3-11 is a means of predicting whether the shot noise or the thermal noise is dominating.

Continuing then, the noise voltage per root Hertz at the filter input is:

$$v_n = i_n \times R = \sqrt{2q\, I_{dc}} \times R = (3.2 \times 10^{-19} \times 3 \times 10^{-9})^{1/2} \times 160 \times 10^6 = \frac{4.9\mu V}{\sqrt{Hz}}$$

and for $\overline{B}$=5 kHz, the total noise voltage is:

$$V_n = v_n \times \sqrt{\overline{B}} = 4.9 \times 10^{-6} \times (5 \times 10^3)^{1/2} = .35 \text{ mv rms.}$$

Notice that we have ignored the thermal noise of R since its contribution is much less ($V_{dc} \gg 50mv$).

We see that bandpass filtering has had no effect on the total noise in the 5kHz band; the noise characteristic is determined by the d.c. leakage current of the diode. To carry this one step further, let us open the input of our receiver to the ambient. Any background light (sunlight, moonlight, etc.) causes an increase in the d.c. diode curent and a corresponding increase in the total noise voltage. Thus, even though we have used a modulated carrier for our light beam, the d.c. levels, in this case background light levels, have an irreducible effect on the system performance.

Although the modulated carrier transmitter-bandpass receiver system does not improve the noise performance in this example, in an actual system, involving active elements, there are other low frequency noise sources and d.c. offset drifts that make a purely d.c. system unattractive.

3.4   ACTIVE DEVICES

In virtually all applications involving low level signal processing, active devices (operational amplifiers, transistors) are used to increase the voltage or power levels. The active device can significantly degrade the noise performance of these systems unless device selection and design rules are followed. Low noise considerations are most appropriate where a transducer, e.g., a microphone, antenna, phone pickup, strain gauge, etc., occurs in a system. Ideal circuit performance is achieved when the ultimate resolution of the system is determined only by the characteristics of the tranducer. Fortunately, transducer limited performance can be achieved in most applications using presently available active elements.

Our focus is on describing the noise characteristics of an operational amplifier. Although the internal devices of the op amp are not discussed in depth, it should be remembered that in a multi-stage high gain circuit, e.g., an operational amplifier, the input stage determines the overall noise performance of the device. Therefore, we expect the noise characteristics of a particular operational amplifier to resemble the characteristics associated with the input stage devices. For example, a field effect transistor input op amp has the general noise properties of a field effect transistor and similarly for bipolar input devices. Section 3.4.6 explores the internal structure of the op amp and its effect on noise performance.

Equivalent Noise Voltage and Noise Current

Two characteristics of the input device determine the op amp noise characteristics: 1) internal resistances-channel resistance in the F.E.T. or base spreading resistance in the bipolar transistor; and 2) bias currents-gate leakage in the F.E.T. and base current in the bipolar transistor. When source resistances are high, the shot noise associated with the d.c. bias or leakage currents of the inputs is the dominant noise component. When the source resistance is low or zero, the thermally generated noise of the internal resistances is dominant. The following measurement circuit (Figure 3-12) serves to define the equivalent noise voltage e_{ni} and noise current i_{ni} by considering the operational amplifier while operating from zero source resistance (e_{ni}) and high source resistance (i_{ni}).

3.4.1   Noise Voltage e_{ni}

Figure 3-12 shows a typical noise voltage test setup. Resistors R_1 and R_2 are required to bias the operational amplifier into its linear region and to accurately fix the voltage gain. R_1 is limited to several hundred ohms to appear as an effective short circuit to the shot noise currents associated with the input bias currents and to effectively eliminate its thermal noise contribution at the input of the operational amplifier. Under these conditions, the output noise voltage is the equivalent noise voltage generated between the inputs multiplied by the gain $(R_1 + R_2)/R_1$ of the circuit. In practice, we measure the rms noise per root Hertz at the output of the amplifier with the measuring circuit of Figure 3-1. We then divide these values by the gain of the circuit $(R_1 + R_2)/R_1$ and form the graph of input

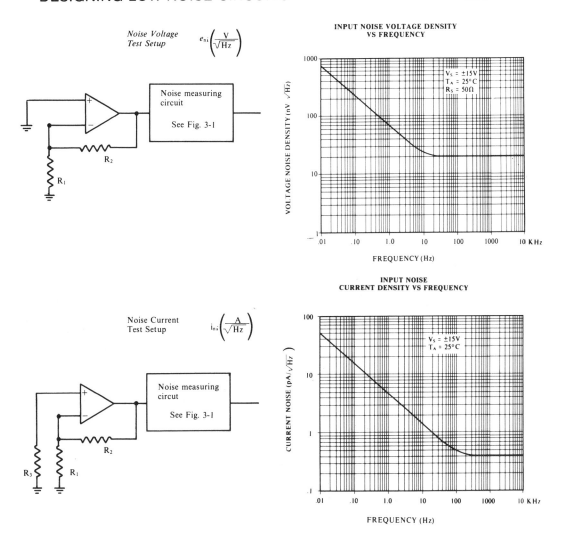

Figure 3-12. Test set-up for noise measurements.

noise voltage versus frequency shown in Figure 3-12. Note that for frequencies above about 100 Hz, this noise spectrum has the white or "flat" noise characteristic which we encountered with thermal and shot noise sources. Starting at about 10Hz, however, the input noise per root Hertz increases with decreasing frequency. This important region is discussed in Section 3.4.4.

This definition of input noise voltage e_{ni} allows us to draw the equivalent circuit of Figure 3-13.

Here we have replaced our noisy operational amplifier with a

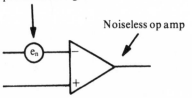

Equivalent input noise voltage source

Noiseless op amp

Figure 3-13. Input noise voltage model.

noiseless amplifier and an input noise source e_{ni}. Because the output of the noise source is random, we may put the source e_{ni} in either the inverting or the noninverting terminal, *but not both*! As we see from the equivalent circuit, nothing we do to the inputs alters the effect of e_{ni} on the noise characteristics of the operational amplifier. This is *not* the case for the equivalent input noise current density, i_{ni}, considered next.

3.4.2 Noise Current i_{ni}

Figure 3-13 shows the test setup for measuring the equivalent input noise current. Except for the flicker noise region (see Section 3.4.4) the input noise current is composed almost entirely of the shot noise contributed by the input bias current I_b. Thus, by making R_3 and R_1 large, we can force the shot noise due to I_b in these resistors to dominate the thermal noise (see Section 3.3.3), and the output noise voltage is generated by noise currents alone. In practice, we may wish to restrict the maximum values of R_1 and R_3, and, of course, the noise voltage e_{ni} is always present. However, both e_{ni} and the resistor noises are known and so we can subtract these factors out and calculate the equivalent input noise current i_{ni}. Figure 3-12 shows a graph of i_{ni} versus frequency for a typical operational amplifier. This noise current flows in both the inverting *and* non-inverting terminals. We can account for this fact with the equivalent circuit of Figure 3-14. Remember that the magnitude of the noise current source in either input is given by the graph of Figure 3-12, but these sources are independent, so the total noise voltage at input due to noise currents flowing in the source resistances in Figure 3-12 is:

$$e_n = \sqrt{(i_{ni} \times R_{eq}^-)^2 + (i_{ni} \times R_{eq}^+)^2} = i_{ni} \sqrt{(R_{eq}^-)^2 + (R_{eq}^+)^2} \qquad (3\text{-}12)$$

where R_{eq}^- = equivalent resistance to ground on the inverting (−) terminal; and R_{eq}^+ = equivalent resistance to ground on the non-inverting (+) terminal.

Equivalent input noise current sources

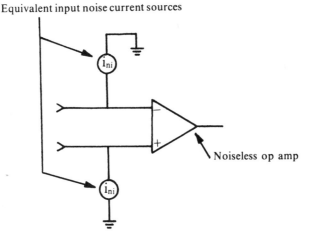

Figure 3-14. Input noise current model.

If we are dealing with mean square values e_{ni}^2 or i_{ni}^2 then the total mean square noise voltage at the input due to the noise current is:

$$\bar{e}_{ni}^2 = \bar{i}_{ni}^2 \times R_{eq}^+ + \bar{i}_{ni}^2 \times R_{eq}^- \qquad \textbf{(3-13)}$$

Since most amplifiers are specified with rms rather than mean square noise parameters, Equation 3-12 is generally more useful.

3.4.3   Complete Noise Model

We can combine our noise voltage and noise current models to obtain the general noise model of Figure 3-15. Again e_{ni} could appear in either input.

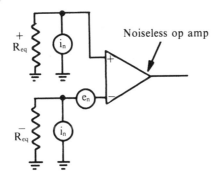

Figure 3-15. Complete input noise model.

This model allows us to calculate the total noise voltage at the inputs due to all sources. R_{eq}^+ and R_{eq}^- result from the transducer or previous stage source resistance and any feedback or biasing resistors that connect to the amplifier's inputs. A complete input noise voltage calculation must include the thermal noise of R_{eq} and R_{eq}^- and the shot noise voltages created by any d.c. currents that might be flowing in these resistors in addition to the input bias currents. The output noise voltage is found by multiplying the total input noise voltage by the stage gain.

By considering each noise source of Figure 3-15, we can write the total input noise density e_{in} (total) as:

$$e_{ni} \text{ (total)} = \left[e_{ni}^2 + (i_{ni} \ R_{eq}^+)^2 + (i_{ni} \ R_{eq}^-)^2 + (e_{nt}^+)^2 + (e_{nt}^-)^2 \right]^{1/2} \quad \textbf{(3-14)}$$

and the total input noise voltage in a given bandwidth is just:

$$E_n \text{ (total)} \ = \ e_n \text{ (total)} \sqrt{f_H - f_L} \quad \textbf{(3-15)}$$

where e_{nt}^+ and e_{nt}^- are the thermal noise contributions of R_{eq}^+ and R_{eq}^-, and f_H and f_L are the upper and lower frequency cutoff points of our system. Equation 3-13 is valid in the white noise region only. For simplicity, we have assumed that no d.c. currents other than input bias currents flow in R_{eq}^+ and R_{eq}^-. Figure 3-15 can be used to calculate the input noise voltage density and is most useful when the transducer is modeled as a voltage source in series with a resistor (e.g., resistive strain gauge). If the source model is a current source paralleled with a resistor (e.g., photo diode), then it is more useful to calculate the total input noise current density i_{in} (total), by converting all noise voltage sources to current sources and summing as above.

Examples: In the following room temperature example, we use the noise model of Figure 3-15 and assume that our operational amplifier has the noise characteristic graphed in Figure 3-12. We restrict our treatment, for the moment, to the flat portion of the noise density graphs.

Example 1: Figure 3-16 is a voltage amplifier with a gain of +21.

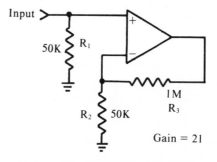

Figure 3-16. Noninverting amplifier.

Assuming no input signal, Part 1, calculate the total input noise voltage between 1000 and 7000 Hz. Part 2, what is the total output noise voltage?

Part 1: We first calculate the equivalent resistor at each input. For the noninverting input, $R_{eq}^{+} = 50k\Omega$. At the inverting input, we have $50k\Omega$ in parallel with $1m\Omega$ thus, $R_{eq}^{-} = 47.6k\Omega$. We must substitute values for e_{ni} and i_{ni} in the equivalent circuit of Figure 3-15. Looking at Figure 3-12, we find that $e_{ni} = 20nV/\sqrt{HZ}$ and $I_{ni} = 4pA/\sqrt{HZ}$ over the frequency range 1000 Hz to 7000 Hz (flat portion of the curves). Using Equation 3-14, we have:

$$\underline{e_n}\,(total) = \left[e_{ni}^2 + (i_{ni}\,R_{eq}^{+})^2 + (i_{ni}\,R_{eq}^{-})^2 + (e_{nt}^{+})^2 + (e_{nt}^{-})^2\right]^{1/2}$$

$$\underline{e_n}\,(total) = \left[(\frac{20nV}{\sqrt{Hz}})^2 + (\frac{.4pA}{\sqrt{Hz}} \times 50K\Omega\,)^2 + (\frac{.4pA}{\sqrt{Hz}} \times 47.6K\Omega)^2 \right.$$

$$\left. + \frac{1.6 \times 10^{-20} \times 50K\Omega}{Hz} + \frac{1.6 \times 10^{-20} \times 47.6K\Omega}{Hz} \right]^{1/2}$$

$$\underline{e_{ni}}\,(total) = \left[4 \times 10^{-16}\,\frac{V^2}{Hz} + 4 \times 10^{-16}\,\frac{V^2}{Hz} + 3.6 \times 10^{-16}\,\frac{V^2}{Hz} \right.$$

$$\left. + 8 \times 10^{-16}\,\frac{V^2}{Hz} + 7.6 \times 10^{-16}\,\frac{V^2}{Hz} \right]^{1/2} \qquad \text{(3-16)}$$

$$\underline{e_{ni}}\,(total) = 52\,\frac{nV}{\sqrt{Hz}}\cdot$$

And for a bandwidth of 6000 Hz we haved from Equation 3-15.

$$\underline{E_n}\,(total) = 52\,\frac{nV}{\sqrt{Hz}} \times \sqrt{6000Hz} = 4.03\mu V \text{ rms.}$$

Part 2: The total output voltage is $E_n\,(total)_{out} = E_n(total) \times$ circuit gain = $4.03\mu V \times 21 = 84.6\,\mu$ Volts.

Looking at the intermediate calculation, (Equation 3-16), we see that all noise sources in this example are of similar magnitudes with the thermal noise of R_{eq}^{+} and R_{eq}^{-} being the largest single contribution. Since we have chosen a typical operational amplifier and reasonable resistor values, we may say that in most designs all noise sources need to be considered.

Since the feedback resistors implicit in R_{eq}^{-} contribute significantly to the total noise in this design, let us consider reducing their values. Recall that we need a gain of 21. Let $R_1 = 500$ and $R_2 = 10k\Omega$. Then $R_{eq} = 476\Omega$ and Equation 3-16 becomes:

$$e_{ni}\,(total) = \left[4 \times 10^{-16}\,\frac{V^2}{Hz} + 4 \times 10^{-16}\,\frac{V^2}{Hz} + \underbrace{3.6 \times 10^{-20}\,\frac{V^2}{Hz}}_{\text{3rd term}} \right.$$

$$\left. + 8 \times 10^{-16}\,\frac{V^2}{Hz} + \underbrace{7.6 \times 10^{-20}\,\frac{V^2}{Hz}}_{\text{5th term}} \right]^{1/2}$$

and we see that the 3rd and 5th terms are greatly reduced in value. Our total noise voltage is now:

$$E_n(\text{total}) \;\;=\;\; 3.0\mu V \text{ rms}$$

which is a 25% reduction in the total noise voltage. R_{eq}^+ is still 50K to maintain high input resistance so we now have unbalanced input resistors and the output voltage offset due to the input bias currents may increase to an unacceptable level.

3.4.4   Flicker Noise

We have seen that the noise spectral densities associated with thermal and shot noise sources are "flat" or white. For frequencies above about 1 kHz, depending on the device, active elements exhibit a white spectral density as well. However, in the region below 1 kHz active devices exhibit a noise density which increases as frequency decreases. In this region the noise is proportional to $1/f$, where f is the frequency, and has been called $1/f$ ("one over f") noise. More recently, the term "flicker noise" has been used and the term "excess noise" is also applied to this effect. The origins of the flicker phenomenon are not entirely understood although it is clear that the surface conditions of the semiconductor wafer are critical. In addition, there is speculation that certain thermodynamic properties of any bulk material produce a $1/f$ resistivity fluctuation due to random equilibrium *thermal currents*. Thus, it is possible that the flicker phenomenon is an essential physical property of matter much like thermal and shot noise. In any case, present manufacturing efforts aimed at improving the semiconductor surface conditions have reduced, but not eliminated, the flicker noise effect.

Obviously, we need to consider the flicker noise contribution only if the frequency response of our system extends below $\simeq 100$ Hz. Furthermore, we can almost always avoid this spectral region by modulation of our transducer, regardless of our system bandwidth (see Section 3.4.8). In addition, this modulation technique avoids most d.c. offset and drift errors and is, therefore, very attractive. Unfortunately, modulation or chopping techniques increase system complexity and may not be cost effective in many moderate performance systems. Therefore, we need a technique for calculating the wideband contribution of the flicker noise in the $1/f$ region.

Figure 3-17 shows the input noise voltage of a low noise operational amplifier, the SSS 725 manufactured by Precision Monolithics, Inc. This graph emphasizes the flicker noise region. As before, the total noise E_n in a given bandwidth (f_H - f_L) is the area under the noise voltage density curve (Figure 3-17) in the bandwidth of interest. In the white noise region, this calculation is a simple multiplication. However, in the flicker noise region, we must integrate the noise density curve over the frequency range of interest to find the area under the curve E_n. To simplify this calculation in the $1/f$ region, we find the $1/f$ noise voltage corner frequency f_{NV} by extrapolating from the white noise and $1/f$ noise curves as shown in Figure 3-17. Along the $1/f$ noise curve the noise voltage density is

$$e_n = e_n \text{ (white) } \sqrt{\frac{f_{NV}}{f}} \qquad \text{(3-17)}$$

for the region $0 < f < f_{NV}$, where e_n (white) is the noise voltage density in the white noise region. To find the area under the curve, we find the integral of equation 3-17 in the frequency band $f_H - f_L$, assuming that f_H and f_L are below the $1/f$ noise corner frequency thus,

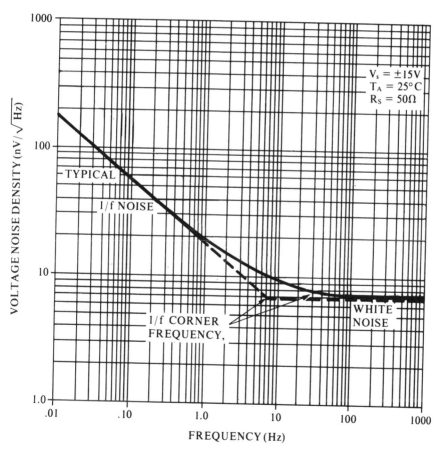

Figure 3-17. Flicker noise region for SSS 725.

$$e_n = \int_{f_L}^{f_H} \frac{e_n}{\sqrt{H_z}} = \frac{e_{n(white)}}{\sqrt{H_z}} \int_{f_L}^{f_H} \sqrt{\frac{f_{NV}}{f_L}} \, df = \frac{e_{n(white)}}{\sqrt{H_z}} \sqrt{f_{NV} \, ln(\frac{f_H}{f_L})} \qquad \text{(3-18)}$$

Example: Using the graph of Figure 3-17, let the 1/f noise corner frequency, f_{NV} = 7 Hz, and e_n (white) = $6.8\frac{nV}{\sqrt{Hz}}$

Part 1: Find the noise voltage density for the SSS 725 operational amplifier at .001 Hz.

Solution: Since .001 Hz is not on the graph, we use equation 3-17. Thus,

$$e_n = e_n (white) \sqrt{\frac{f_{NV}}{f}} = 6.8 \frac{nV}{\sqrt{Hz}} \sqrt{\frac{7 \ Hz}{.001 \ Hz}} = .56 \frac{\mu V}{\sqrt{Hz}}$$

Part 2: What is the rms noise voltage in the bandwidth .01 Hz to 1 Hz.
Solution: Using Equation 3-18 we have:

$$E_n = e_n (white) \sqrt{f_{NV} \, ln (\frac{f_H}{f_L})} = 6.8 \frac{nV}{\sqrt{Hz}} \sqrt{7 \ Hz \, ln \frac{1}{.01}} = 38 \ nV \ rms.$$

Similarly, we can write equations for the noise current density i_n and for the noise current I_n in the bandwidth (f_H - f_L). Equation 3-17 is replaced with:

$$i_n = i_n (white) \sqrt{\frac{f_{NC}}{f}} \frac{A}{\sqrt{Hz}} \ \text{for} \ 0 < f < f_{NC}$$

and equation 3-18 is replaced with :

$$\text{(3-19)}$$

$$I_n = i_n (white) \sqrt{f_{NC} \, ln (\frac{f_H}{f_L})} \ A \ (rms)$$
$$0 < f < f_{NC}$$

Remember that the 1/f corner frequency of the noise current, f_{NC}, is usually different from f_{NV}.

It should be noted that the 1/f frequency dependence of flicker noise is an empirical relation which has been extensively varified in both semiconductors *and* resistive elements. Fortunately, except for certain types of resistors (e.g., carbon composition), the flicker noise magnitude is very low in most resistors and for good quality metal film or wirewound resistors the effect can be ignored.

3.4.5   General Input Noise Calculation

We may bring the effects of flicker and white noise together in an equation for the total input noise voltage and total input noise current. Thus,

$$e_n = e_n \text{ (white)} \sqrt{\frac{f_{NV}}{f} + 1} \; \frac{V}{\sqrt{Hz}} \text{ at frequency f} \qquad \text{(3-20)}$$

and

$$E_n (f_H - f_L) = e_n(\text{white}) \sqrt{f_{NV} \; ln \; \frac{f_H}{f_L} + (f_H - f_L)} \; V \text{ (rms)} \qquad \text{(3-21)}$$

Further,

$$i_n = i_n \text{ (white)} \sqrt{\frac{f_{NC}}{f} + 1} \; \frac{A}{\sqrt{Hz}} \text{ at frequency f} \qquad \text{(3-22)}$$

and

$$I_n (f_H - f_L) = i_n \text{ (white)} \sqrt{f_{NC} \; ln \; (\frac{f_H}{f_L}) + (f_H - f_L)} \; A \text{ (rms)} \qquad \text{(3-23)}$$

Note that Equations 3-20 through 3-23 are valid even in the vicinity of the noise corner frequency.

Example: Use Equation 3-20 to find the noise voltage density of the 725 amplifier at the 1/f corner frequency. Assume

$$e_n(\text{white}) = 6.8 \; \frac{nV}{Hz} \text{ and } f_{NV} = 7 \text{ Hz}$$

Solution:

$$e_n = 6.8 \; \frac{nV}{\sqrt{Hz}} \sqrt{\frac{7 \text{ Hz}}{7 \text{ Hz}} + 1} = 9.6 \; \frac{nV}{Hz} \text{ at 7 Hz}$$

in agreement with the graphed value of Figure 3-17.

For the equivalent circuit of Figure 3-15 we may now write an expression for the total input noise voltage due to all sources. Using Equations 3-21 and 3-23 we get:

$$E_n (f_H - f_L) = \left[e_n^2 (\text{white}) \left(f_{NV} \; ln \; \frac{f_H}{f_L} + (f_H - f_L) \right) + i_n^2(\text{white}) \times \right.$$
$$\left. \left(f_{NC} \; ln \; (\frac{f_H}{f_L}) + (f_H - f_L) \right) (R_{eq}^{-2} + R_{eq}^{+2}) + e_{nt+}^2 + e_{nt-}^2 \right]^{1/2} V(\text{rms})$$

where e_{nt+} and e_{nt-} are the thermal noise contributions of R_{eq+} and R_{eq-}, respectively, over the specified bandwidth. This is a general expression and should be used when the frequency pass band extends into the flicker noise region. Otherwise, Equation 3-14 is adequate.

3.4.6 Selecting the Best Low Noise Operational Amplifier

Our efforts so far have been directed at calculating the total circuit noise for a given operational amplifier and its supporting circuitry. The first step in a low noise design, however, is the selection of the operational amplifier itself. Typically we are presented with a source, e.g. (phone pickup), and its equivalent circuit, and are required to amplify its output (voltage or power) while introducing the least possible noise and harmonic distortion. Any source may be modeled by a voltage source with a series resistor Figure 3-18(a) or a current source with a parallel resistor Figure 3-18(b). Of course, for a given source either representation may be used, but we normally choose the equivalent circuit which most closely represents the device physically e.g., current source for a photo diode: voltage source for a phono pickup. Real sources i.e. ($R_s{\neq}0$, $R_p{\neq}0$), generate thermal noise due to R_s and R_p, and since many transducers require excitation or biasing, shot noises may be significant. The task of the circuit designer, then, is to ensure that the inherent source (transducer) noises are the limiting noise factors. For the majority of sources this level of performance can be closely approached using widely available operational amplifiers. To aid in selecting the best operational amplifier for a given transducer, we introduce the concepts "Noise Factor" and "Noise Figure."

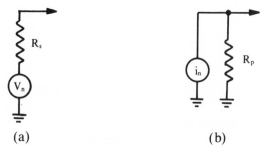

(a)   (b)

Figure 3-18. Transducer Models.
(a) Voltage source. (b) Current source.

Noise Factor and Noise Figure

A commonly used measure of amplifier noise performance is its noise figure or noise factor. The noise factor F is defined by:

$$F = \frac{\text{all noise (mean square at the input)}}{\text{noise due to source (mean square at the input)}}$$

This definition of noise factor is the most useful when dealing with input-referred noise as we have been doing, but many equivalent definitions exist. Note that F is defined in terms of noise power and we must use mean square values not rms values. The noise figure NF is defined by:

$$NF = 10 \log F = 10 \log \frac{\text{all noise (mean square)}}{\text{noise due to source (mean square)}} \text{ decibels}$$

and is more widely used. A perfect amplifier would have:

F   =   1 and
NF =   0 decibels. Typical N.F.'s will be in the range .1-3 decibels (dB)

For the circuit of Figure 3-19 we may write the noise figure as:

$$NF = 10 \log \frac{\text{all noise}}{\text{source noise}} = 10 \log \frac{(e_{ni}^2 + i_{ni}^2 R_s^2) + 4k\,TR_s}{4kTR_s} \qquad \textbf{(3-24)}$$

Where $e_{ni}^2 + i_{ni}^2 R_s^2$ is the amplifier input noise and $4K\,TR_s$ is the source noise.

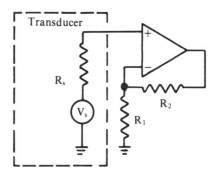

Figure 3-19. Transducer preamplifier.

Equation 3-24 assumes that we have chosen the feedback resistors, R_1 and R_2, so that their noise contribution (thermal and $i_{ni}\,\overline{R_{eq}}$) is negligible. The numerator is the total noise at the input-amplifier and source; the denominator, the source noise alone. Had the source included shot noise components, we would have converted them into mean square noise voltages (i_n shot $R_s)^2$ and added this effect to the numerator and denominator.

Best circuit performance occurs when the N.F. is a minimum, but for constant temperature, the only parameters which we can control are e_n and i_n. Mathematically, it can be shown that Equation 3-24 has a minimum value when:

$$\frac{e_{ni}}{i_{ni}} = R_s, \text{ i.e., when } e_{ni} \text{ is equal to the noise voltage across} \qquad \textbf{(3-25)}$$

R_s caused by i_{ni} ($e_{ni} = i_{ni}R_s$) and the noise figure under these conditions is:

$$N.F._{min} = 10 \log \frac{e_{ni}i_{ni}}{2kT} + 1 \qquad (3\text{-}26)$$

These results are useful in the selection of the best operational amplifier. For a particular operational amplifier at a specified frequency, Equation 3-25 states that an optimum source resistance exists and that the N.F. at that source resistance is given by Equation 3-26. However, there is no guarantee that another operational amplifier with different e_n and i_n characteristics and a different optimum source resistance does not perform better than the original operational amplifier; even though the second operational amplifier is operated with non-optimum source resistance. Thus, the selection process is more involved than simply finding an operational amplifier whose optimum source resistance is close to the R_s, of the transducer. Fortunately, a very general selection technique using noise figures exists.

Operational Amplifier Selection Based on N.F. vs. R_s Curves

Figure 3-20(a) and (b) are the e_{ni} and i_{ni} curves for the SSS 725 operational amplifier (Precision Monolithics, Inc.), a high accuracy instrumentation amplifier specified for low noise application. By choosing values of e_n and i_n at specific frequencies (1Hz, 10Hz, etc.) and using these values in Equation 3-24 we can graph Equation 3-24 as a function of source resistance and generate the family of curves of Figure 3-20(c). From the graph, we can find the N.F. for various source resistances and frequencies. Also, we can find the optimum source resistance at a given frequency. For 1Hz and 10kHz, the noise figures for a given source resistance are poorer than those for higher frequencies due to the $1/f$ noise at low frequencies. Also 100 kHz performance is slightly poorer than 1 kHz performance due to the increase in noise current that begins at $\approx$9kHz. Of course, if both e_n and i_n were white (flat) then all the curves of Figure 2-30(c) would be identical. We see then that $1/f$ noise is a severe performance limitation. For reference, recall that a 3dB noise figure means equal source and amplifier noise *power* and that at 1dB N.F. the amplifier noise power is about 25% of the source noise power. For rms voltage, 3dB N.F. implies that the amplifier noise voltage is 40% of the source noise voltage. For most systems, a 3dB N.F. is quite acceptable.

Using the N.F. vs. source resistance curve of several different amplifiers, we can choose the best device for a given source resistance and frequency. Several manufacturers publish these curves, but by using

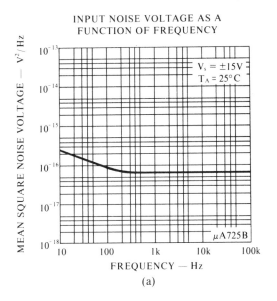

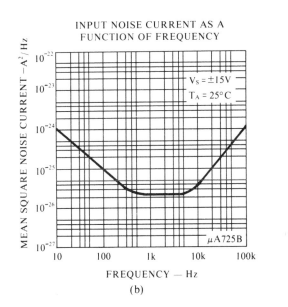

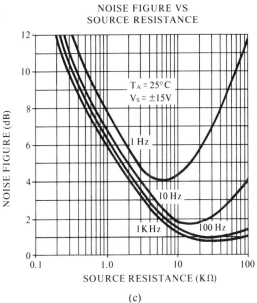

Figure 3-20. Typical input noise characteristics for the SSS 725.

Equation 3-24 and the e_{ni} i_{ni} vs. frequency specs, we can generate these curves
for any operational amplifier.

The noise figure we have been using so far is called the *spot* noise
figure because we are using values of $\bar{e}_n^2$ and $\bar{i}_n^2$ at a specified frequency.
Using Equation 3-24, we can also calculate the wideband N.F. by simply

substituting the broadband input noise parameters $\bar{E}_{ni}^2$ and $\bar{I}_{ni}^2$ for $\bar{e}_{ni}^2$ and $\bar{i}_{ni}^2$ using the broad band source noise $\bar{E}_{ni}^2 = 4k\,TR_sB$. Equation 3-24 becomes:

$$N.F. = 10\log\frac{\bar{E}_{ni}^2 + \bar{I}_{ni}^2\,R_s^2 + 4kTR_s\,\bar{B}}{4kTR_s\,\bar{B}} \qquad (3\text{-}27)$$

Notice that in the white noise region where:

$$\bar{E}_{ni}^2 = \bar{e}_{ni}^2 \times \bar{B} \text{ and } \bar{I}_{ni}^2 = \bar{i}_{ni}^2 \times B \text{ equation 3-27 becomes}$$

$$N.F. = 10\log\frac{\bar{e}_n^2\,\cancel{B} + \bar{i}_{ni}^2\,\cancel{B}\,R_s^2 + 4kTR_s\,\cancel{B}}{4kTR_s\,\cancel{B}} = \frac{\bar{e}_{ni}^2 + \bar{i}_{ni}^2\,R_s^2 + 4kTR_s}{4kTR_s}$$

which says that the spot and broadband N.F. are equal in the white noise region of the $\bar{e}_{ni}^2$ and $\bar{i}_{ni}^2$ curves. Thus, Equation 3-27 is used only if the passband extends into the 1/f region. Under these conditions $\bar{E}_{ni}^2$ and $\bar{I}_{ni}^2$ are found using Equations 3-21 and 3-23. The wide band N.F. graph for a bandwidth ($f_H - f_L$) always lies between the two spot noise graphs for f_H and f_L for a given amplifier.

Example: Compare the noise performance of the SSS 725 bipolar input operational amplifier with the LH0042 (National Semiconductor). FET input operational amplifier, at various source resistances. The amplifier is used in a 10 Hz bandwidth centered at 100 Hz.

Part 1: Since the application is narrowband about 100 Hz we need only compare the spot noise figures of the two amplifiers at 100 Hz. Figure 3-20(c) has just the graph we need to specify the SSS 725 performance, but we find National Semiconductor does not publish such a graph for its LH0042. The National data sheet does list the e_{ni} and i_{ni} at 100 Hz as $55nV/\sqrt{Hz}$ and $2.2fA/\sqrt{Hz}$ respectively. Thus, the noise factor of the LH0052 as a function of R_s at 100 Hz becomes, from Equation 3-24:

$$N.F. = 10\log\frac{(3.0\times10^{-15} + 5.0\times10^{-30}R_s^2) + 1.6\times10^{-20}R_s}{1.6\times10^{-20}R_s}$$

which we can simplify to:

$$N.F. = 10\log\left[\frac{3(10^{-15}) + R_s^2(5)(10^{-30})}{1.6(10^{-20})R_s} + 1\right] \qquad (3\text{-}28)$$

Now, we plot a few points of this equation for various source resistances R_s. As a starting point, we find the optimum source resistance R_s (opt) from Equation 3-25 thus:

$$R_s(opt) = \frac{e_{ni}}{i_{ni}} = 2.5M \text{ and from Equation 3-26 the N.F. is:}$$

$$N.F._{(min)} = 10 \log \left[\frac{e_{ni} \quad i_{ni}}{2kT} + 1 \right] = .065dB @ R_s = 2.5M$$

The remainder of the curve is generated by choosing resistance values about R_s (opt) and solving Equation 3-28. Both N.F. curves appear in Figure 3-21. As a performance reference the N.F. of the popular 741 operational amplifier appears also.

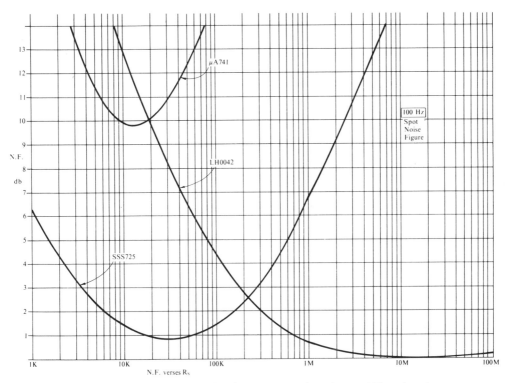

Figure 3-21. Choosing the best operational amplifier for low noise applications.

We see that for $R_s <$ 200kΩ the SSS 725 functions better, and for $R_s >$ 200Ω the LH0042 is our choice. In the region around 200 kΩ other factors such as price, d.c. drift specifications, bandwidth, etc. influence our choice. Clearly, the 741, even when operated at its optimum source resistance ($\sim$14kΩ), is an inferior choice.

General Selection Criteria

We have developed a rigorous technique for determining the suitability of any device in a particular low noise application. Ideally, using only the e_{ni} and i_{ni} versus frequency curves and given an R_s and bandwidth requirement, we could make a graph similar to Figure 3-21 which includes all low noise candidates. This group can be quite large; so we should restrict our analysis to parts whose general characteristics recommend them. Some of these general noise characteristics are outlined below.

In our example. we purposely chose to compare an FET and bipolar input type. As we saw, their individual ranges of application are quite different and we shall consider FET input devices and bipolar input devices separately.

Bipolar

The bipolar input part (e.g., SSS 725) is most appropriate where source resistances are on the order of 100 kΩ or less. These devices have the advantage of very low d.c. offsets and minimum parameter variation over temperature. For a given bipolar transistor, the values of e_{ni} and i_{ni} are strong functions of collector current. Since e_{ni} decreases and i_{ni} increases as the collector current is increased, we find that the value of $R_{s(opt)}$ decreases under these conditions. This implies that operational amplifiers designed for low power applications tend to have larger $R_{s(opt)}$ than general purpose or wideband types. Similarly, low input bias current parts, inasmuch as this implies reduced collector currents, have larger $R_{s(opt)}$. In general, bipolar parts are used if the transducer is modeled by a voltage source ($R_s < 100$ kΩ) and for lowest noise under these conditions the noninverting configuration is required. Estimation of i_{ni} using the d.c. input bias current, I_{dc}, (i.e. $i_{ni} = \sqrt{2q\ I_{dc}}$) may yield values that are too low and hence is unreliable for bipolar devices. Refer to manufacturer's data sheets for i_{ni} specifications.

FET Input Types

These components function well in high impedance (> 100kΩ) applications. Unfortunately, the d.c. performance (offsets and drifts) of the FET is comparatively poor unless source resistances are quite high (>10 MΩ). Many FET input data sheets list only e_{ni} noise specifications, since measurement of i_n is difficult and a reliable estimation of i_{ni} can be generated by applying the shot noise equation, $i_{ni} = \sqrt{2q\ I_{dc}}$ where I_{dc} is the published leakage specs for either input. Since i_{ni} is caused exclusively by input leakage currents and since the input leakage doubles with every 10°C increase in *chip* temperature, i_{ni} and $R_{s(opt)} = e_{ni}/i_{ni}$, varies accordingly. Many FET input types exhibit very low ($<.1$ db) Noise Figures in the vicinity of their optimum source resistances. This implies excellent performance in high impedance applications such as photo multipliers and photo diodes. Under these conditions the amplifier is always used in its inverting mode (current to voltage). Special consideration should be given to FET types for wideband applications from medium (20 kHz) to high (500 kHz) frequencies since the noise performance of the FET input stage is not degraded by the higher source currents required for high frequency operation. The flicker noise voltage and current corner frequencies are typically an order of magnitude lower than for bipolar types, and for medium source resistance applications ($\sim$100kΩ) at low frequencies ($<$100 Hz) an FET component is a more attractive choice (versus bipolar) than it was at medium frequencies for identical source conditions.

When choosing either type it is frequently helpful to see the internal circuit diagrams of the operational amplifier. These diagrams are almost always supplied by the device manufacturer. There are several things to check. Best noise performance is achieved by the common emitter (common source) configurations preferably with resistive, not active, loads in the collectors (drains). For high frequencies, however, the larger input capacitances associated with this configuration (Miller effect) may degrade the input impedance sufficiently to affect the N.F. (see Section 3.3.7). Under these conditions emitter follower (source follower) or cascade input configurations are preferred. Most low noise preamplifiers have large voltage gains and if low distortion must be maintained over a wide frequency range, then the gain bandwidth of the component must be high, and therefore uncompensated components are preferred over internally (unity gain) compensated components.

3.4.7   Amplifier Input Resistance and Capacitance

The input resistance of most operational amplifiers is 1M or greater open loop, and if feedback is applied in the noninverting mode, this figure could increase to 100M or more. By using a resistor in parallel with the input we could set the input resistance to any lower value. We might ask if there is an optimum input resistance that "matches" our noise source. Let us examine the circuit of Figure 3-22.

Here we have a voltage source and a series resistor and we have added the parallel input resistor R_{in} to a noiseless amplifier. If we write the noise figure, we have, using mean square values,

$$\text{N.F.} = 10 \log \frac{\text{all noise}}{\text{source noise}} = 10 \log \frac{\text{Thermal noise of } R_s \text{ and } R_{in}}{\text{Thermal noise of } R_s} \qquad \textbf{(3-29)}$$

The numerator of this expression is $4KT (R_s R / R_s + R_{in})$, but the denominator is more complicated. The source noise is $4KT R_s$, but because of the voltage divider, R_s and R_{in}, all of this noise does not appear at the amplifier input. Further, we are dealing with mean square noise voltage, and

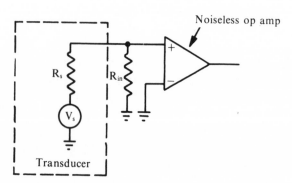

Figure 3-22. Transducer preamp with finite input resistance.

so the source noise *at the input* becomes $(R_{in}/R_s + R_{in})^2\, 4kTR_s$. Thus, Equation 3-29 becomes:

$$\text{N.F.} = 10 \log \frac{4kT \dfrac{R_s}{R_s + R_{in}}\dfrac{R_{in}}{}}{\dfrac{R_{in}^2}{R_s + R_{in}}\, 4kTR_s} = 10 \log \frac{R_s + r_{in}}{R_{in}} = 10 \log \left[1 + \frac{R_s}{R_{in}}\right]$$

This is the noise figure of a perfect amplifier with finite input resistance.

We see immediately that the lowest N.F. occurs when $R_{in} \gg R_s$. This may come as a surprise to some who, guided by power matching requirements, might have chosen $R_{in} = R_s$. The same calculation could be performed for a current source working into the inverting input and here we find that $R_{in} = 0$ yields the lowest noise figure. This requirement is met by the virtual ground at the summing mode of an inverting amplifier.

Input Capacitance

The noise figure for the perfect amplifier of Figure 3-23 with a capacitive input is:

$$\text{N.F.} = 10 \log \frac{\text{all noise}}{\text{source noise}} = 10 \log \frac{4kT\,[\text{Re}Z]}{4kT\,[\text{Re}Z]} = 0 \text{ where}$$

ReZ is the real part of the complex impedance at the amplifier input.

Thus, for a *perfect amplifier*, the input capacitance does not affect the noise figure. Notice, however, that both the noise and the signal are reduced equally as we increase frequency because the impedance of the capacitor is decreasing. With a *real amplifier,* the input noise voltage e_{ni} is unchanged by the input conditions and thus the noise figure is degraded at higher frequencies because the ratio of input noise to source noise is

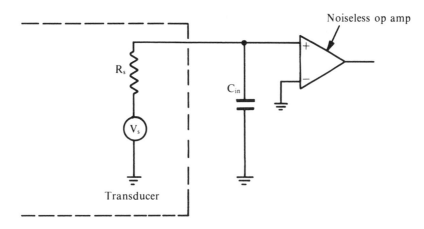

Figure 3-23. Transducer preamp with capacitive input.

increasing. Note, however, that the published (input to input) capacitance of a device is reduced if feedback is applied and the feedback network is of relatively low resistance.

3.4.8 Other Techniques

Modulation. We may avoid the 1/f noise spectrum in low frequency applications by modulating the transducer if the particular transducer type allows it. Consider the system of Figure 3-24.

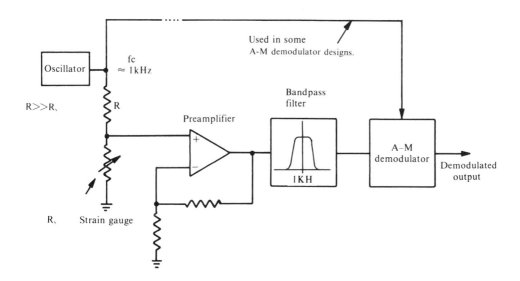

Figure 3-24. Modulated transducer preamplifier.

In this system, the low frequency variations in the strain gauge become amplitude variations at the carrier frequency. Thus, the low frequency (f_m) information is placed in side bands ($f_c - f_m$) and ($f_c + f_m$) about the carrier. The bandpass filter must have a passband $>2f_m$. A further advantage of this system is the elimination of d.c. offsets and drifts associated with the preamplifier.

Transformer. As we saw in the discussion of the Noise Figure, the optimum source resistance for most operational amplifiers is greater than 5 kΩ. Using an operational amplifier directly, it is difficult to maintain a low noise figure when operating from low source impedances ($<$5k Ω). Since an ideal transformer is a noiseless element, it can be used to transform a low source impedance to a higher value and thus "match" a particular operational amplifier. Ideally, an extremely low noise figure ($<$0.1db) could be obtained in this fashion by matching our source to a good quality FET amp. In practice this is difficult to do with a real transformer. Core loss and wire resistance limit the transformer performance and for low frequencies the primary inductance must be kept high while for higher frequencies stray capacitance must be limited. Furthermore, the transformer is susceptible to stray electric and magnetic fields.

The transformer finds greatest application at medium frequencies (1 kHz-500 kHz) while transforming low ($<$5 kΩ) source impedance to medium (5 kΩ-50 kΩ) values. Good performance is easier to achieve in narrow band, rather than wideband, applications.

3.4.9 Real Time Noise Representation

We have considered a noise source as containing a specified amount of energy in a given frequency range. This is a frequency-domain representation and it is useful because, for random sources, we are unable to predict the actual voltage or current at a specified point in time. However, we can predict the percentage of time that the voltage or current spends above a given level. Table I is true for the Gaussian noise sources we normally deal with and is expressed as a fraction of the rms value:

Normalized Value (peak/rms)	*Percent of Time Normalized Value is Exceeded*
.675	50%
1	31.7%
2	4.54%
3	.26
4	.0064

Table I

Chapter 4

Oscillators and
Waveform Generators

4.1   A good test oscillator or signal generator is a laboratory must. However, when we need a special waveform or signal in a dedicated test station or in a portable piece of equipment, we abandon our laboratory generator and take to the drawing board. These signals may take the form of square waves, pulses, ramps, sine waves or some other arbitrary periodic wave shape. We may want to vary the amplitude and frequency of these wave shapes and, depending on the application, stringent requirements may be placed on some parameters such as frequency stability, linearity or harmonic distortion products. For example, the oscillator used in a digital wristwatch must not drift more than 0.002% per month for 1-minute accuracy over that time period, whereas the sine wave used to test a low distortion audio preamplifier must have very low distortion.

Several manufacturers have introduced monolithic waveform generators and in circuits where the full capability of these parts can be utilized their expense is justified. However, circuits built with operational amplifiers and comparators are more versatile and usually less expensive when the circuit requirements are not precisely met by a monolithic part.

The circuit of the oscillator-generator takes different forms depending on the application. However, most circuits are based on one of two approaches: the free-running multivibrator or the sine wave oscillator. The output of the multivibrator is a square wave or a string of pulses, but with suitable waveshaping techniques, triangular, sawtooth or even sine waveshapes can be achieved. The sine wave oscillator produces a sine wave directly, generally of lower distortion than the sine wave derived from the

141

multivibrator. We shall first describe a simple free-running multivibrator and a simple sine wave oscillator using operational amplifiers. These basic blocks are then used in various applications.

4.2   FREE-RUNNING MULTIVIBRATORS

The circuit of Figure 4a shows a gated 1 kHz multivibrator using a general purpose operational amplifier, e.g., 741. If the multivibrator is gated "on" by opening switch S_1, the output is a series of square waves (Figure 1b).

The operation can be understood by considering switch S_1 *closed* and $\pm$ 15 volts applied to the supply terminals of the operational amplifier. The resistive dividing network consisting of resistors R_1 and R_2 attached to the noninverting input causes *positive* feedback, and with V_1 at zero potential, any disturbance (noise, voltage offset) at the + input terminal is amplified and fed back in phase causing the output to "latch" at one of the supply voltages. For example, if the output V_o has a positive value at a given instant, a portion $[R_2/(R_1+R_2)V_{o_1}$ of this positive signal appears at the noninverting terminal causing the output to go more positive. This action continues until the output reaches the positive saturation voltage of the amplifier. This is a stable state. If the output is latched at the positive supply voltage, and switch S_1 is opened, then V_1 begins rising at a rate determined by I_T and C_T. When this voltage slightly exceeds V_2, the output starts going negative, and since C_T momentarily "fixes" V_1, the regenerative action between V_o and V_2 causes the output to latch at the negative supply voltage. With the change in output states, I_T is now flowing out of the capacitor towards the output terminal through R_1, and V_1 decreases until it is slightly less than V_2, at which time the output switches to the positive supply voltage. The sequence repeats itself resulting in free-running switching. If S_1 is excluded, circuit operation begins when the supply voltages are applied, although the first few cycles may not be of the correct frequency or waveshape.

4.2.1   Choosing the Timing Components

Approximate Equations. The frequency of operation, f is obtained by the calculation of the time interval t_1 (Figure 1b). For symmetric output voltage waveforms,

$$f = \frac{1}{2t_1.}$$

(4-1)

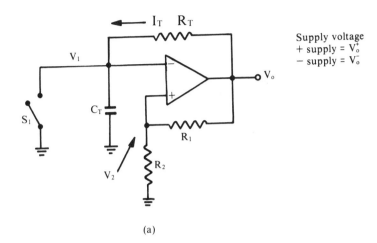

(a)

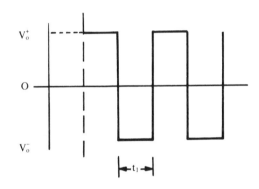

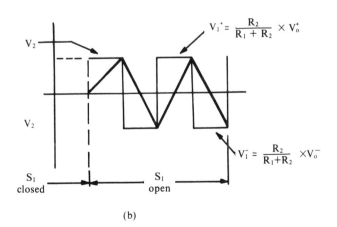

(b)

Figure 4-1. Free running multivibrator.

(a) Circuit diagram
(b) Waveforms.

If we choose R_1 and R_2 so that V_o is much greater than V_2, then I_T is approximately constant over the timing period because the voltage excursion of V_1 is very small relative to V_o, allowing us to write:

$$I_T = \frac{V_o - V_1}{R_T} \approx \frac{V_o}{R_T} \quad \text{for } V_o >> V_1 \qquad \text{(4-2)}$$

At the beginning of the timing period (immediately following switching) I_T is slightly larger than the value given by Equation 4-2 and it is slightly smaller at the end of the timing period, but our approximation is very good for small voltage excursions of V_1 about zero.

Noting that for a capacitor the Charge Q is related to the capacitance C and the voltage V by Q=CV, and that the change in charge, ΔQ, is related to the current I and the time increment Δt by $\Delta Q = I\Delta t$, we find that, for constant charging current, the time required to change the terminal voltage of capacitor C_T by ΔV (volts is:

$$\Delta t = \frac{\Delta V C_T}{I_T} \qquad \text{(4-3)}$$

But in the present case (see Figure 4-1) $\Delta t = t_1$, $V = 2V_o R_2/(R_1 + R_2)$ and $I_T = V_o/R_T$, so:

$$t_1 = \frac{2R_2}{R_1 + R_2} \times R_T C_T$$

and from Equation (4-1):

$$f = \frac{1}{2t_1} = \frac{1}{4}\left(\frac{R_1 + R_2}{R_2}\right) \times \frac{1}{R_T C_T} \quad \text{for } R_1 >> R_2 \qquad \text{(4-4)}$$

Notice that the output voltage cancelled out when Equations 4-2 and 4-3 were combined. Thus, the frequency is independent of the supply voltage.

Looking ahead to the results of the next section, we can estimate the range of accuracy of the above equation. For $R_1 > 10R_2$ better than 1% accuracy is achieved and Equation 4-4 is useful up to $R_1 = R_2$ where the predicted frequency will be 10% too high.

Selection of R_1, R_2, R_T and C_T. It is clear that the combination of the selected values must satisfy Equation (4-4) for the desired frequency, but certain limits must be put on the individual component values. The product $R_T C_T$ must be large with respect to t_1 ($R_T C_T \geq 10t_1$) for the "I_T constant" approximation to hold. It is desirable to make R_T large to allow for a relatively small value of C_T to keep capacitor size and cost down. The upper limit of R_T is determined by the requirement that $I_T \approx V_o/R_T$ be much greater than the maximum input bias current, I_b, of the operational amplifier or comparator to make the operation practically independent of I_b. The values of R_1 and R_2 are chosen so that $V_1 = [R_2/(R_1 + R_2)] V_o \ll V_o$, say $V_1 \approx 0.1 V_o$, leading to $R_2 \approx 0.1 R_1$, while keeping $V_1 \gg V_{offset\ max}$, say $V_1 \approx 10 V_{offset\ max}$ so that variations in V_{offset} from unit to unit and variations with temperature have little effect on the characteristics of the multivibrator. For similar reasons, the resistance $R_1 + R_2$ must be small enough so that the current through $R_1 + R_2$ is much greater than $I_{b\ max}$ (say, approximately $100\ I_{b\ max}$).

For a numerical example we note that for the 741 operational $V_{offset\ max} = 5mV$ and $V_{sat} = 12V$. Thus, we choose $R_T = 12V/(100 \times 500 \times 10^{-9})$ = 240k ohms. This is an EIA standard carbon composition resistor. Similarly, we choose $R_1 = 240k$ ohms, and $R_2 = 0.1 R_1 = 24K$ ohms. From Equation 4-4 for f = 1k Hz, we find $C_T = 0.01$ μF. A standard 0.01 amplifier $Ib_{max} = 500$ μA, 16 volt ceramic capacitor is suitable. The power dissipation (V^2/R) in the resistors is extremely smally so $\frac{1}{4}$-watt carbon film resistors are adequate. Typical tolerances for capacitors are 20% and for resistors $\pm 10\%$ or $\pm 5\%$. Thus, to obtain an exact frequency, it is necessary to incorporate an adjustable component into the circuit. For example, R_2 can be made a standard 0.5-watt 50k ohm potentiometer.

In this example, the frequency of operation is independent of the supply voltage only if the operational amplifier output stage has very low saturation voltages in the positive and negative directions. Unfortunately an operational amplifier like the 741 in saturation has almost a volt of difference between its output and the supply voltage under light loading conditions (>100kohm), and as much as 3 volts difference with a heavy load (<10kohm). Therefore, the multivibrator running frequency is load and device dependent. If very stable frequency is a circuit requirement, choose a comparator, e.g., LM311 which has well defined output states or use an operational amplifier similar to the RCA CA3130 whose MOS output stage very closely approaches the supply voltage.

Exact Timing Equations. For low frequency operation, it is desirable to allow the input voltages to span a greater percentage of the output voltage range. This results in longer timing periods for a given capacitor-resistor combination. In this case, however, the timing equations are no longer simple since the current I_T cannot be considered constant.

In Figure 4-2 the resistive voltage dividing network R_1, R_2 and the timing diagram illustrate the critical voltages in the positive and negative states. Let us consider the period t_1 in Figure 4-2. The output has just switched to a positive value (V_o^+) and C_T begins charging from V_2^- toward V_2^+ through resistor R_T. Introducing the voltages V_x and V_y measured with reference to V_2^- as shown in Figure 4-2a and 4-2b, we find that for $V_y = 0$ at t = 0:

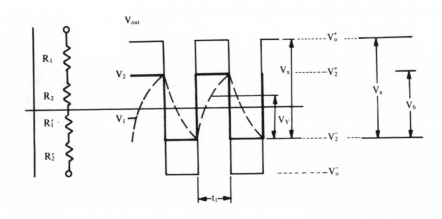

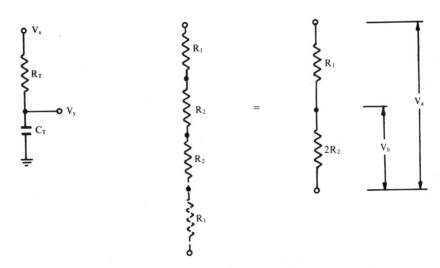

Figure 4-2. Diagrams for exact multivibrator frequency computation.

(a) Voltages in relation to regenerative circuit
(b) Regenerative circuit equivalence.

$$\frac{V_y}{V_x} = 1 - \exp\left(-\frac{t}{R_T C_T}\right)$$

which expresses the fraction of the total supply voltage V_x which V_t reaches as a function of time for a given set of R_T and C_T. The output of the multivibrator switches states when the fraction V_y/V_x is equal to V_b/V_a as shown in Figure 4-2. But V_b is the output of the resistive divider network formed by R_1, R_2 and the reflection of R_2 in the negative direction with V_a as the input. Thus, with the aid of Figure 4-2c we can write:

$$\frac{V_b}{V_a} = \frac{2R_2}{R_1 + 2R_2}$$

Switching occurs when $V_y/V_x = V_b/V_a$ so t_1 can be found by solving:

$$1 - \exp\left(\frac{-t_1}{R_T C_T}\right) = \frac{2R_2}{R_1 + 2R_2}$$

Rearranging terms and taking the natural logarithm of both sides we get:

$$t_1 = R_T C_T \ln\left(1 - \frac{2R_2}{R_1 + 2R_2}\right) \tag{4-5}$$

where t_1 is the time the multivibrator spends in either state.

Figure 4-3 is a graph of $\ln(1 - 2R_2/(R_1 + 2R_2))$ as a function of R_2/R_1. For example, if $R_1 = R_2 = 10K$, then $R_2/R_1 = 1$ and from the graph $\ln(1 - 2R_2/R_1\, 2R_2) = 1.098$. From Equation 4-5:

$$t_1 = R_T C_T \times 1.098.$$

As before, the frequency f is given by:

$$f = \frac{1}{2t_1}$$

and if for example, $R_T = 100$kohm and $C_T = .1\mu f$ then:

$$t_1 = 10^5 \times 0.1\mu f \times 1.098 = 1.098 \text{ms}.$$

The frequency operation is:

$$f = \frac{1}{2t_1} = 455 \text{ Hz}$$

The selection of components is done as was shown in the previous paragraphs, except the restriction $R_2 = 0.1R_1$ is removed, and Equation 4-4 is replaced with Equation 4-5 or Figure 4-3.

4.2.2   Choosing an I.C. for Multivibrator Application

For best performance, observe the following precautions:
1) The amplifier must be able to withstand large differential input

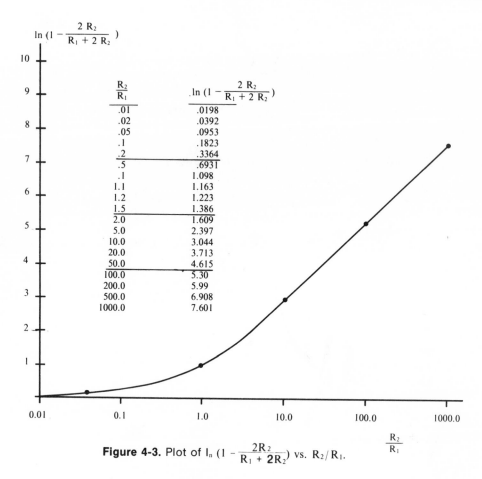

Figure 4-3. Plot of $I_n (1 - \frac{2R_2}{R_1 + 2R_2})$ vs. R_2/R_1.

voltages without increasing the input currents. Immediately after the output changes state (see Figure 4-1b) there is several volts difference between the two inputs. Since large differential voltages are not encountered in linear applications, many operational amplifiers specified for linear operation do not function correctly in multivibrator circuits. For example, some FET and super beta input types conduct current when the *differential* input voltage exceeds several hundred millivolts due to diode protection networks connected between the inputs. This obviously violates the assumption that negligible current flows into the inputs. Avoid operational amplifiers which have this type of input protection.

2) For frequencies above about 5kHz, general purpose operational amplifiers and even some high speed types do not function well in multivibrator circuits. Although these amplifiers may have rise time specifications that would imply operation above 5kHz, the problem is poor recovery from saturation. The large input differentials (see Figure 4-

1b) cause all the amplifier stages to saturate and when the switching point is reached all of these stages must recover from saturation before the output transition can occur. For example, a 741 is specified for a $1\mu s$ rise time, but $10-15\mu s$ are required for the amplifier to recover from excessive saturation.

To alleviate this problem voltage comparators must be used since they are designed and specified for saturated switching performance. Care must be taken in the application of voltage comparators, however, since their output voltage levels are usually not as convenient as operational amplifiers which switch essentially between the plus and minus supply levels. Operation above 10 MHz is possible with fast comparators.

3) Input bias currents contribute errors, especially in very low frequency applications where the timing resistor is large. These errors generally take the form of duty cycle variations, since the bias current flows always in the same direction, while the charging current reverses direction when the output voltage reverses polarity. Choose an I.C. which has specified input bias currents significantly lower than the charging current which flows through the timing resistor. Polarized (electrolytic) capacitors can be used if the negative terminal is returned to the appropriate supply. This results in a significant size reduction for a given capacitor value. Furthermore, a smaller valued timing resistor can be used and an amplifier with larger bias currents can be tolerated.

4.3   SINE WAVE OSCILLATORS

While multivibrator circuits use an amplifier or comparator as a switch, a sine wave oscillator utilizes the operational amplifier in its linear mode. Addition of frequency selective positive feedback results in a controlled frequency instability. This is normally done by connecting the *input* of a filter that selects a given frequency to the *output* of an operational amplifier and feeding the *output* of the filter to the noninverting *input* of the amplifier. If the filter has zero phase shift at frequency f_o, then we have positive feedback around the amplifier and oscillations begin at f_o. The amplitude of these oscillations increases until some limiting behavior decreases the forward gain of the amplifier to a value that just compensates for the filter attenuation at f_o. In the uncontrolled situation this limiting occurs when the output saturates at the supply voltages resulting in a square-wave-like shape. However, the amplitude can be stabilized at lower values by controlling the gain of the amplifier with an external network.

In the circuit of Figure 4-4, R_1 and R_2 form the resistive divider used to set the gain of the operational amplifier in the conventional manner. The Wien bridge filter network attached to the noninverting input selects frequency f_o as shown in Figure 4-4b. If the gain as determined by R_1 and R_2

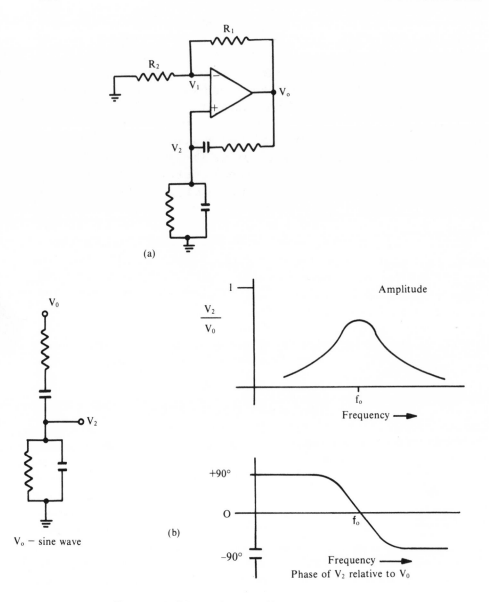

Figure 4-4. Wien bridge oscillator.

(a) Circuit diagram
(b) Characteristics.

exactly compensates for the filter attenuation at f_o, then stable oscillations result. Note that if the operational amplifier gain is slightly larger than necessary, the output begins clipping (saturating at the supply voltages) thus limiting the gain. If the operational amplifier gain is slightly less than required, the circuit ceases to oscillate.

To aid in the understanding of this circuit, consider breaking the feedback loop of an oscillator running at frequency f_o as shown in Figure 4-5, and driving point 1 with a variable sine wave generator. If we view the signal generator waveform at point 1 and the amplifier output (point 2) on a dual trace oscilloscope as we vary the generator signal, at one frequency f_o the amplitude and phase of the signal at point 2 is identical to the test signal. At all other frequencies the amplitude at point 2 is less than the test signal amplitude and the phase either leads or lags.

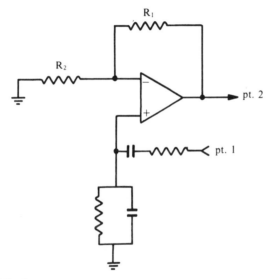

Figure 4-5. Opening the oscillator loop helps to illustrate the proper conditions for stable oscillations.

The circuit of Figure 4-6 is one of the simplest *stabilized* sine wave oscillators. Amplifier gain control is achieved by the output voltage sensitive feedback network attached to the inverting input. Diodes D_1 and D_2 cause the gain of the amplifier to decrease as the out signal amplitude increases. The degree of this effect can be controlled by R_{adj}. If R_{adj} is several times larger than R_1, the forward gain at zero signal level is determined by R_1 and R_2, but as the signal amplitude increases D_1 or D_2 allows R_{adj} to appear in parallel with R_1 causing the gain to decrease. Since R_{adj} is larger than R_1, the gain reduction has a gradual, smooth effect. In practice, gain reduction of only a few percent below the zero signal level gain is required.

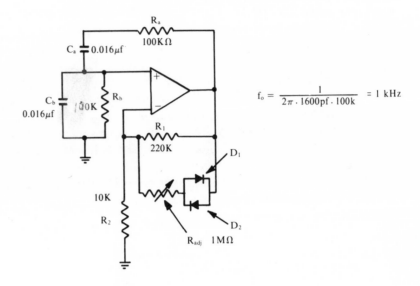

$$f_o = \frac{1}{2\pi \cdot 1600 \text{pf} \cdot 100\text{k}} = 1 \text{ kHz}$$

Figure 4-6. Practical sinewave oscillator

Zero phase shift frequency for any R & C.	Zero phase shift frequency for $R_aC_a=R_bC_b$
$f_o = \dfrac{1}{2\pi \sqrt{R_aR_bCC_b}}$	$f_o = \dfrac{1}{2\pi R_aC_a}$
Attenuation at f_o:	Attenuation at f_o:
$B = \dfrac{1}{1 + \dfrac{R_a + C_b}{R_b = C_a}}$	$B = \dfrac{1}{1 + 2\dfrac{R_a}{R_b}}$

Table I

Let us analyze the oscillator of Figure 4-6. With the component values shown in the Wien bridge, zero phase shift and therefore oscillations occur at about 1kHz.

Since $R_aC_a = R_bC_b$, we can use the simplified expression for the bridge attenuation, B given in Table I. With $R_a = 100\text{K}$ and $R_b = 10\text{K}$:

$$B = \frac{1}{21}$$

Thus, the forward gain at zero signal level must be slightly larger than 21 for oscillations to begin. With $R_1 = 220K$ and $R_2 = 10K$ the forward gain at zero signal level is 23. R_{adj} is chosen to cause the gain to decrease to exactly 21 at the required signal amplitude.

Although the circuit of Figure 4-6 is useful because of its simplicity, it suffers from several limitations:

1) An adjustable resistor is almost always required if the output amplitude is tightly specified.

2) Because of the nonlinear forward gain characteristics an amount of distortion is essential to this circuit. However, for the oscillator shown, measured distortion is less than 1% at 1kHz. Distortion can be reduced by increasing the value of R_{adj} relative to R_1 and changing R_2 so that the zero signal gain is closer to 21. This results in a "softer" nonlinear characteristic. Unfortunately, temperature sensitivity is increased causing amplitude variations.

A circuit which eliminates these problems is presented in Section 4.4.3.

4.3.1   Choosing an I.C. for Sine Wave Oscillator Applications

Since the operational amplifier in a sine wave oscillator is used as a linear amplifier, the usual considerations described in Chapter 1 are relevent. However, special consideration should be given to :

1) *Common-Mode Rejection Ratio.* The operational amplifier of Figure 4-4 or Figure 4-6 is used in a noninverting configuration and thus a fraction of the output signal forms a common-mode signal for the inputs. This fraction is exactly the attenuation ratio through the feedback network R_1 and R_2. In an effort to limit distortion due to nonlinearities in the operational amplifier, we would like to set the forward gain to a relatively small value, say 10, with resistors R_1 and R_2. However, as we continue to decrease the gain, the common mode signal is increasing and eventually the distortion increases due to inadequate common mode rejection, especially at higher frequencies. Therefore, restrict the common mode voltage on the inputs to no more than a few volts.

2) *Frequency Responses.* Even though the amplifier is oscillating, we cannot ignore the possiblity of unwanted oscillations. For uncompensated operational amplifiers, follow the manufacturer's recommendation for frequency compensation for the gain as determined by R_1 and R_2. Choose an operational amplifier for higher frequency applications much as you would choose an amplifier for that frequency. In general, the open-loop gain should be at least 10 times the closed-loop gain at the oscillation frequency. This must include the gain reducing effects of frequency compenstion networks.

4.4 APPLICATIONS

The oscillator circuits just discussed form the basis for many generator applications. Several of these are treated in the following sections.

4.4.1 Triangular Wave Generation

The circuit shown in Figure 4-7 generates both an accurate large amplitude triangular wave and a square wave using two operational amplifiers. Circuit operation is similar to that of the multivibrator of Figure 1a, but the timing feedback connection is made to the noninverting input instead of to the inverting input, because of the signal inversion through the integrator (operational amplifier No. 2).

Assuming that operational amplifier No. 1 has just switched to its high level, we see from the timing diagrams that the triangle output is also high (instant t_o). Thus, the voltages at point a and b are high and because of the positive feedback connection, operational amplifier No. 1 is latched high. The output voltage of the integrator heads downward and, pulled via the resistive divider R_1 and R_2, the + input of amplifier No. 1 is driven in a negative direction. When the voltage at point c reaches ground level (at instant t_1), operation amplifier No. 1 latches at the low voltage and the integrator output starts charging toward the positive level again. Operational amplifier No. 1 returns to the positive latch state when point c crosses zero in the positive direction (t_2).

Assuming similar operational amplifiers, R_2 must be smaller than R_1 to allow the output of the integrator to pull the + input of operational amplifier No. 1 across zero.

The charging current into or out of the timing capacitor is a constant determined by the output saturation voltage (V_{sat}) of operational amplifier No. 1 and R_T. The slew rate ($\Delta V / \Delta t$) of the triangle waved is

$$\frac{\Delta V}{\Delta t} = \frac{I_T}{C_T} = \frac{V_{sat}}{R_T C_T}$$

Assuming symmetric + and – saturation voltages for operational amplifier No. 1, the operating frequency is:

$$f = \frac{R_1}{4R_2} \times \frac{1}{R_T C_T} \quad \text{for } R_1 > R_2 \qquad \textbf{(4-6)}$$

This equation is exact and useful for all values of R_1 and R_2 provided $R_1 > R_2$.

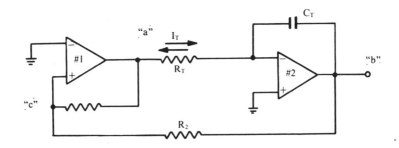

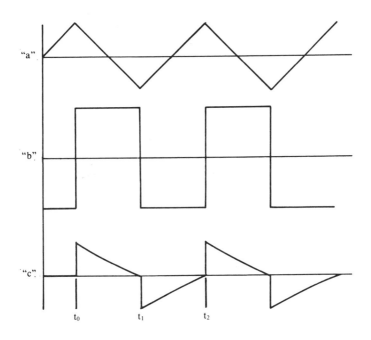

Figure 4-7. Triangular waveshape oscillator.

(a) Block diagram
(b) Waveshapes.

Note that there is only a slight difference between Equation 4-6 and Equation 4-4.

Although a wave form similar to a true triangle shape is generated at the negative input of the simple multivibrator of Figure 4-1a, if we try to increase the amplitude by adjusting R_1 and R_2, the wave shape becomes distorted because the charging current into C_T is no longer constant over the timing period. However, in the circuit just presented the output is a true integrator and linearity is maintained up the the full output range of operational amplifier No. 2 and, the output is a low impedance source.

4.4.2 True Triangle Waves Find Application in Many Systems

Figure 4-8 shows a voltage controlled duty cycle modulator. If the control voltage, V_C, is zero, then the output V_{out} is a symmetric square wave. If V_C is greater than zero, the output spends proportionately less time in the high state and vice versa for V_C less than zero. Since the output is a switching

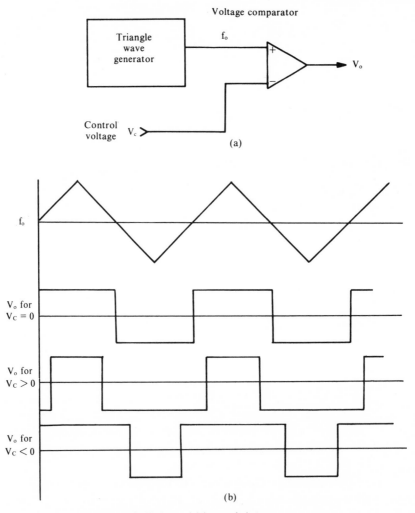

Figure 4-8. Pulse width modulator.

(a) Block diagram. (b) Waveforms

function, it is readily adapted to pulse or digital transmission systems such as optical communication. The high degree of linearity inherent in this type of triangle wave generation guarantees a high fidelity pulse width modulation communication channel. At the receiving end, a simple low pass filter is all that is needed to recover the modulated information (Figure 4-9).

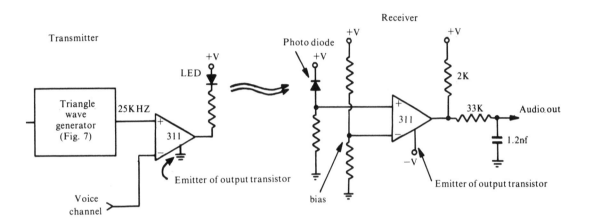

Figure 4-9. Pulse width modulator communications channel.

4.4.3   High Performance Sine Wave Oscillator

Figure 4-10 illustrates a very accurate method of stabilizing the basic Wien bridge oscillator (see Section 4.3) using a field effect transistor (FET) as a variable resistor in the negative feedback loop of the oscillator I.C. For small drain to source voltages (<100mv) an FET can be considered a variable resistor (see Figure 4-11) whose value is controlled by the gate to source voltage (V_{gs}). The lowest value of resistance occurs at $V_{GS}= 0$ and for an n-channel FET the resistance increases for gate voltages more negative than

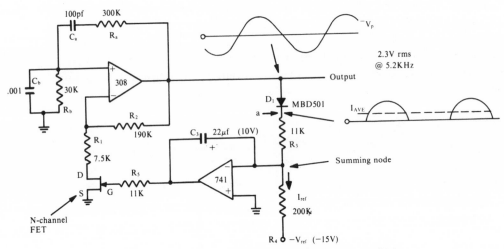

Figure 4-10. High performance sine wave oscillator.

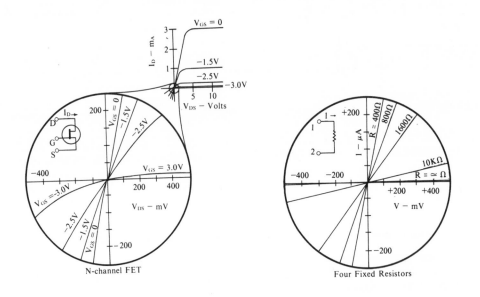

Figure 4-11. Comparison of FET and resistor characteristics.

the source voltage. Notice that the drain can be either positive or negative relative to the source if we restrict ourselves to small voltage excursions. In the circuit of Figure 4-10 with 2.3V rms output, the voltage across the FET is about 30mv rms and, therefore, the FET approximates a variable resistor quite closely. Since the FET is in series with R_1, increasing the resistance in this leg by lowering V_{gs}, reduces the gain of the oscillator I.C. and vice versa.

In this design, the 741 acts as a current comparator. Capacitor C_3 averages the half-wave current pulses which appear at the summing node due to D_1 and R_3 and compares this positive current, I_{ave}, with the negative reference current from R_4, I_{ref}. If I_{ave} is slightly larger than I_{ref}, the output of the 741 goes more negative, which reduces the oscillator gain and therefore the output amplitude, thus lowering I_{ave} until $I_{ave} = I_{ref}$. For I_{ave} less than I_{ref} the reverse occurs; a stable output amplitude proportional to the reference voltage is obtained. The current comparison is extremely accurate since the full open-loop gain of the 741 is used for the d.c. currents. The feedback capacitor C_3 simply integrates the half-sine-wave current from D_1 yielding a low ripple d.c. control voltage at the FET gate.

Choosing the Circuit Components

a) *Oscillator I.C. and Wien Bridge.* The Wien bridge components are chosen as before (Section 4.3). For $R_aC_a = R_bC_c$, $R_a = 10R_b$ and with a frequency of 5 kHz, we have:

$$R_aC_a \;=\; \frac{1}{2\pi f}\; \frac{1}{2\pi(5000)} \;=\; 31\,\mu sec.$$

Choosing the convenient capacitor values of 100nf and .001 μf results in $R_3 = $ 300 kohm and $R_b = $ 30 kohm. For these values, the filter attenuation is:

$$B \;=\; \frac{1}{1+2\,\dfrac{R_a}{R_b}}\; \frac{1}{21}$$

Assuming an FET resistance of 3 kohm, the values of R_2 and R_1 shown in Figure 4-10 result in an amplifier gain of about 21 for normal conditions. To optimize results, do not use a unity gain compensated amplifier such as the 741 for the oscillator I.C. since it has insufficient open-loop gain at 5kHz to minimize distortion products due to amplifier nonlinearities. For example, the LM308 amplifier compensated for a gain of 21 has an open-loop gain three times greater than the 741 at 5 kHz, providing improved performance.

b) *FET.* N-channel, P-channel, insulated gate, enhancement, depletion, etc.; almost any field effect transistor works. Most commonly, a junction N or P channel component is used because they are cheap and readily available. For junction FET's, the minimum resistance occurs at zero gate voltage; for P-channel types the equivalent resistance increases for positive gate to source voltage and for N-channel FET's the opposite is true. The N-channel FET used in this design has approximately 600ohm drain to source resistance at zero gate voltage. In operation the gate voltage is about −3 volts and the channel resistance is 2kohm. Best linearity is obtained for channel resistances in the low kilohm range for most FET's. The resistors R_1 and R_2 should be chosen to yield the desired voltage gain at the nominal operating resistance of the FET and maintain the peak drain voltage below $\pm$ 100 mv.

c) *FET Control Circuitry.* Diode D_1 is a half-wave rectifier which feeds output amplitude information to the comparing network. For half-sine-wave input, the average current through R_3 will be:

$$I_{ave} \; = \; \frac{I_P}{\pi} \; = \; \frac{V_P}{R_3\pi}$$

where I_p and V_p are the peak values of current and voltage, respectively. When V_p is known, we can choose R_4 and the negative reference voltage $-V_{ref}$ by setting I_{ref} equal to I_{ave}. Thus:

$$I_{ref} \; = \; \frac{V_{ref}}{R_4} \; = \; I_{ave} \; = \; \frac{V_P}{R_3\pi}$$

The resistors R_3 and R_4 should be as large as possible without introducing errors due to the input bias current requirements of the comparing amplifier.

The resistor and capacitor, R_3 and C_3, form a single pole filter operating on the voltage wave form at point a. If C_3 is too small, then half-wave ripple appears at the FET gate causing resistance modulation in the FET channel, creating harmonic distortion products in the oscillator output due to gain variations during the cycle. The attenuation of this filter should be sufficient to reduce the ripple at the FET gate to about 1% of the drain to source peak-to-peak voltage. The attenuation of this network is approximately:

$$B \; = \; \frac{Re(Z_c)}{R_3} \; = \; \frac{\frac{1}{2\pi fC}}{R_3} \; = \; \frac{1}{2\pi fCR_3}$$

For the circuit of Figure 4-10, the gate ripple is about 300μV peak-to-peak.

Should the output of the comparing amplifier ever go positive during power turn-on or turn-off, resistor R_5 prevents excessive current flow into the gate of the FET. Any value between about 1K and 1Meg is satisfactory.

Performance Limitations

The performance of a fixed frequency sine wave oscillator is determined by three parameters: frequency stability, output amplitude stability and harmonic distortion content. Usually, we are interested in how temperature affects these variables, but supply voltage and time dependencies may also be important.

*Frequency Stability.*Frequency stability is determined by the stability of the passive components in the frequency selective feedback network and by the stability of the phase shift through the operational amplifier. Of these two effects, the operational amplifier phase shift is the more difficult to control since we can purchase passive components whose temperature dependencies are controlled within several parts per million per degree celsius (ppm/C°). Even though the operational amplifier is used with a negative feedback loop (R_1 and R_2, Figure 4-10, for example), the closed-loop phase shift is only reduced from the open-loop value by a fraction equal to the ratio of closed-loop gain to open-loop gain. At higher frequencies the open-loop phase shift is high and gain low and therefore the phase shift observed under feedback conditions could be several degrees. Since the open-loop gain and phase shift are temperature-and-supply-voltage dependent, the closed-loop phase shift is not predictable.

To understand how phase shifts can affect frequency stability and to point the way to a solution, consider the oscillator of Figure 4-10 and the Wien bridge characteristics shown in Figure 4-4b. If the oscillator I.C. has zero phase shift, then the frequency of oscillation is exactly f_o since this satisfies the oscillation criterion, i.e., zero-loop phase shift. However, if the I.C. has 3° of phase *lag* then oscillations do not occur at f_o, but at some *lower* frequency where the filter contributes 3° of phase lead, because +3° –3° = zero phase shift. Similarly, if the I.C. has phase lead, oscillations occur at some higher frequency in this example. Therefore, if the I.C. phase shift is sensitive to some variable, e.g., temperature, the output frequency is temperature-sensitive as well. Unfortunately, the Wien bridge filter is not very frequency selective (low Q) and a few degrees of phase shift result in a rather large frequency deviation. However, for highly selective filters (high Q) a large change in phase is created by a very small frequency shift near the resonant frequency. Therefore, large amplifier phase shifts are compensated by small frequency changes allowing us to relax our specifications on amplifier phase shift sensitivity. A crystal meets this high Q requirement and an oscillator circuit using a quartz-crystal is treated in Section 4.5. High Q inductors are available for a broad range of frequencies and, in conjunction with a suitable capacitor, a high Q resonant tank can also be used to supply the highly selective frequency feedback necessary for stable operation.

Amplitude Stability. The primary limitation on amplitude control is the sensing of the amplitude itself. The sensing circuit must rectify and filter the output to obtain a d.c. level which is proportional to the amplitude of the output. In the circuit of Figure 4-10 the rectifying diode is the limiting element since its voltage drop is temperature sensitive. In this design a Schottky barrier diode was used because of its low forward voltage drop. This represents an improvement over p-n junction devices but for maximum accuracy an active precision half-wave or full-wave rectifier should be used (See Chapter 6, Section 6.2.1). Of course, the reference current and comparing operational amplifier must have d.c. performance specs. consistent with the desired accuracy.

Harmonic Distortion Products. In the simple Wien bridge circuit of Figure 4-10, best performance is achieved if the peak voltage across the FET is kept small. Furthermore, a good operational amplifier with high open-loop gain at the resonant frequency should be used. However, for ultra-low distortion applications a different approach is used. The trick is to use the frequency selective filter as both a selective feedback element *and* as a filter for the output wave form. In the circuit of Figure 4-12, for example, the L-C resonant tank shunts all output signals other than the resonant frequency to ground. For best results, R_o should be made equal to the parallel resistance of the tank at resonance. Under these conditions, half of the amplifier output is returned to the amplifier input only at the resonant frequency. All other frequencies, including harmonics, are attenuated by the filter and R_o. This approach suffers from having an output impedance equal to $\frac{1}{2} R_o$, but a precision voltage follower, as shown in Figure 4-12, can overcome this limitation and still maintain low distortion.

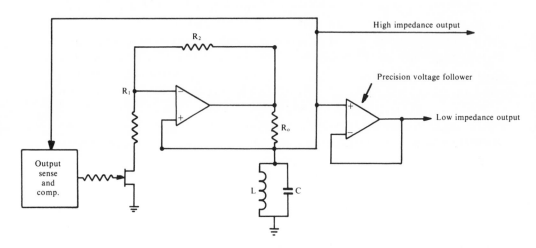

Figure 4-12. Low distortion oscillator using an L-C tank in parallel resonance.

4.5 QUARTZ CRYSTAL OSCILLATORS

A combination of high Q (greater than 10,000) and very low center frequency drift with temperature ($\approx 10 \text{ppm} / \text{C}°$) and time ($\approx 2 \text{ppm} / \text{year}$) makes a quartz crystal oscillator the best choice when extremely accurate frequency stability is required. For example, a low-cost, crystal oscillator-controlled digital watch which loses only .01 seconds/day is easily constructed.

A "crystal" is basically a small chunk of quartz with vacuum-deposited metal electrodes applied to selected faces. The resulting two-terminal device is electromechanically resonant. The equivalent electrical circuit is shown in Figure 4-13 with typical circuit values for a 32 kHz crystal. The values of L_c and C_c are somewhat surprising. L_c results from the crystal mass; C_c is its mechanical compliance (elasticity); R_s is the combined effect of all mechanical losses, including any energy lost to entrapped gas molecules, and C_s is the combination of stray header capacitance and the capacitance of the metalized quartz structure itself.

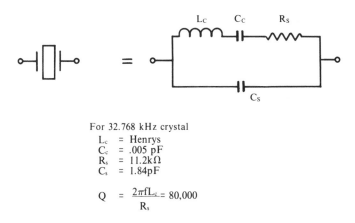

For 32.768 kHz crystal

L_c = Henrys
C_c = .005 pF
R_s = 11.2kΩ
C_s = 1.84pF

$$Q = \frac{2\pi f L_c}{R_s} = 80,000$$

Figure 4-13. Crystal equivalent circuit (series resonance).

Two resonant modes are possible: series resonance and parallel resonance (anti-resonance). In the series resonant mode, the reactances of L_c and C_c cancel, leaving R in parallel with C_s. This is the low resistance mode. Redrawing the equivalent circuit slightly aids in understanding the parallel resonant mode (see Figure 4-14). The series resistance R_s is shown as its equivalent parallel resistance R_p. The capacitively tapped resonant tank consists of L_c in parallel with the series combination of C_c and C_s. Since C_s is so much larger than C_c, the equivalent capacitance appearing in parallel with L_c is only slightly *smaller* than C_c and so parallel resonance occurs at a very slightly *higher* frequency than series resonance. At resonance, the tap point (crystal terminals) appears as a large resistor whose value is given by:

$$R_{in} = \frac{C_c}{C_s} \times R_s Q^2 \qquad (4\text{-}7)$$

In contrast to the series resonant mode, there is no parallel capacitance at

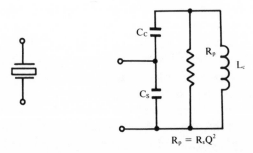

$$R_p = R_s Q^2$$

Figure 4-14. Crystal equivalent circuit (parallel resonance).

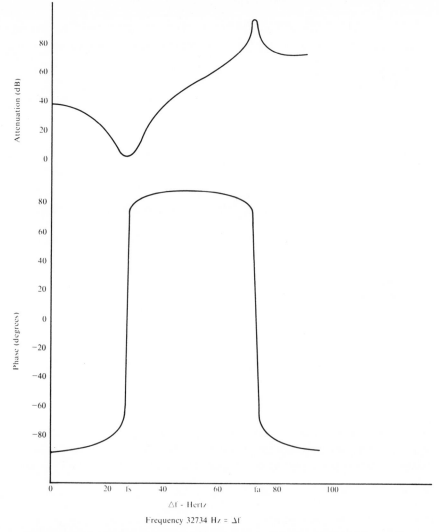

Figure 4-15. Crystal attenuation and phase behavior near resonance.

anti-resonance. Figure 4-15 illustrates the attenuation and phase behavior at the two resonant frequencies.

The following circuit uses the crystal in the series resonant mode. Figure 4-16 shows a medium speed comparator operated as a crystal controlled oscillator. In this design the floating emitter of the 311 comparator is connected to the negative supply, and the open collector load resistor is returned to the positive supply. Resistor R_1 supplies 100% negative feedback for d.c. signals, biasing the comparator into its linear region. This ensures that oscillations begin. Capacitor C_1 removes the negative feedback at the oscillation frequency. Although the output is a square wave, the high selectivity of the crystal causes a sine wave at the series resonant frequency to appear at the noninverting terminal of the 311. Resistor R_2 should be about 10% of the crystal series resistance. The RC filter composed of R_1 and C_1 should attenuate the resonant frequency by a factor of at least 100.

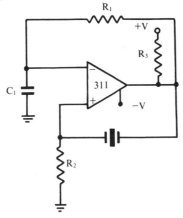

Figure 4-16. Series resonant crystal oscillator using a voltage comparator.

Although the Q_s of crystals are very high, the actual values may vary by a factor of 4 for identically manufactured components. This is due to variations in the series resistance R_s. Therefore, it is difficult to predict the signal attenuation in a particular circuit. This means that sine wave output oscillators are difficult to achieve using crystals since the forward and reverse gains cannot be easily balanced for the different crystals in a production run. However, if a bit of tuning is allowed, most of the previously discussed sine wave oscillators can be used with crystals. A low drift oscillator of high spectral purity results.

Variable Frequency Crystal Oscillators. If the comparator in Figure 4-17 had no switching delay, this circuit would oscillate at exactly the series resonant frequency of the crystal since the combination of zero phase delays

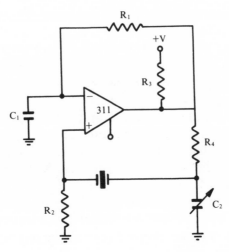

Figure 4-17. Variable frequency crystal oscillator (series resonance).

through both crystal and comparator satisfies the oscillation criterion. If, however, a fixed phase *lead* or *lag* is introduced in the loop, the circuit will oscillate at a slightly different frequency to allow the crystal to supply the required *lag* or *lead* respectively. Unfortunately, the very rapid phase changes associated with small frequency deviations in the crystal limit the range of adjustment using this technique. For example, using the crystal of Figure 4-13 we could expect a maximum frequency shift of about 5 Hertz. Figure 4-18 shows a practical variable frequency crystal oscillator. Resistor

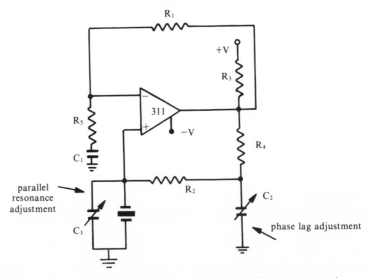

Figure 4-18. Variable frequency oscillator (parallel resonance).

R_4 and capacitor C_2 are chosen to provide a variable phase lag near the series resonant frequency of the crystal. Obviously, the particular crystal should not load the lag network and sufficient loop gain must be maintained. To preserve zero loop phase, the crystal will operate slightly above resonance. The output frequency will increase in response to an increase in C_2.

Using the crystal in the parallel resonant mode permits a wider range of variability. In this mode, resonance is determined by the capacitance C_p in parallel with the equivalent crystal inductance L_c. From Figure 4-14 we see that C_p is the series combination of C_c and C_s. Thus:

$$\frac{1}{C_p} = \frac{1}{C_c} + \frac{1}{C_s}$$

By adding a variable capacitor across the crystal terminals we may add to the value of C_s. Because C_s is much larger than C_c, this additional capacitance causes a relatively small increase in C_p, but a significant increase in tuning range results since the *actual anti-resonant of the crystal* frequency is changed. The most useful circuit for anti-resonant crystal operation is the *crystal π network*. This circuit provides a 180° phase shift at resonance and has, therefore, the advantage of using a simple inverter stage as the gain element. To understand the phase relationships in the network, consider the circuit of Figure 4-19(a). Recall that at anti-resonance, the crystal is a capacitively tapped resonant tank which appears resistive. Therefore, in Figure 4-19(b) points a and b are in phase at anti-resonance. In Figure 4-19(b) capacitor C_v has been replaced with an equivalent series combination. Again, points a and b have the same relationship to each other as in Figure 4-19(a), but because of the ground location points, b and c are 180° out of phase relative to ground. By simply redrawing Figure 4-19(b) and replacing the crystal equivalent circuit with its schematic representation, we have the familiar crystal network, Figure 4-19(c). At resonance, b and c have their maximum values, but b is in phase with a, and c is 180° out of phase. Furthermore, this structure can provide voltage gain; if the output capacitor is *smaller* than the input capacitor the voltage signal at c will be larger than that at b and vice versa. Specifically:

$$\frac{V_{out}}{V_{in}} = -\frac{C_{in}}{C_{out}} \quad \text{the minus sign indicates phase inversion} \qquad \textbf{(4-8)}$$

Figure 4-20 shows a crystal oscillator using the network. In this design a CMOS digital gate biased into its linear region by R_1 is used as an inverter providing 180° phase shift. The crystal π network provides 180° additional phase shift so the total loop shift is zero (360°) and oscillations occur at the parallel resonant (anti-resonant) frequency. The general design equations for the crystal π network are readily developed. The input and output resistance at points b and c, respectively, relative to ground can be

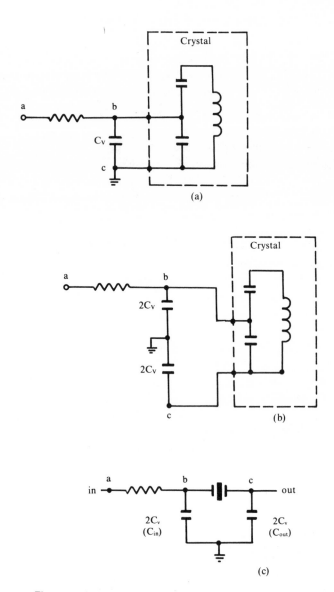

Figure 4-19. Understanding the crystal pi network.

found by first calculating the resistance between points b and c at resonance. Since Figure 4-14 and Figure 4-19 are equivalent, Equation 4-7 is the necessary formula for R_s. However, C_s in Figure 4-14 has an additional parallel capacitance added to it, namely the series combination of C_{in} and C_{out}. Therefore Equation 4-7 becomes:

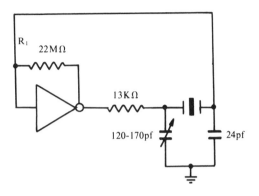

crystal R_s = 1.35K
L_c = 27.7 H
C_c = .0117 pF
C_s = 6.18 pF

279.611 kHz crystal
ocillator using a CMOS inverter

Figure 4-20. 279.611 KHz crystal oscillator using a CMOS inverter.

$$R_{bc} = \left(\frac{C_c}{\frac{C_{in} \times C_{out}}{C_s\, C_{in} + C_{out}}} \right)^2 \times R_s Q^2 \qquad (4\text{-}9)$$

With the π tapping network, the input resistance R_b becomes

$$R_b = R_{bc} \left(\frac{C_{out}}{C_{in} + C_{out}} \right)^2 \qquad (4\text{-}10)$$

And the output resistance R_c is:

$$R_b = R_{bc} \left(\frac{C_{in}}{C_{out} + C_{in}} \right) \qquad (4\text{-}11)$$

Equation 4-10 and Equation 4-11 satisfy power conservation requirements in the tank and therefore $R_c + R_b$ is not equal to R_{bc}. The input and output voltage relationships are given in Equation 4-8.

4.6   OSCILLATORS FOR DIGITAL SYSTEMS

Many digital systems require a system clock for timing synchronous operations or simple sequencing. If a gate of the chosen logic family (e.g., TTL or CMOS) can be used as the gain element in the oscillator circuit, then no special circuits (that may need additional power supplies) are required and the oscillator output will have system compatible logic levels. Since the basic gate of any logic family must have significant power gain, this approach is always feasible. The circuit of the preceding example, Figure 4-20, illustrates such an oscillator. Several other oscillators that use digital gates are discussed below.

4.6.1   Gateable R.C. Controlled Oscillator Using CMOS Gates

Figure 4-21(a) shows a digitally controlled multivibrator type oscillator using CMOS nand gates operated from symmetric + and – 5 volt supplies. For simplicity assume for the moment that the gates switch at the middle of the supply voltage range, that is, zero (0) volts in this example.

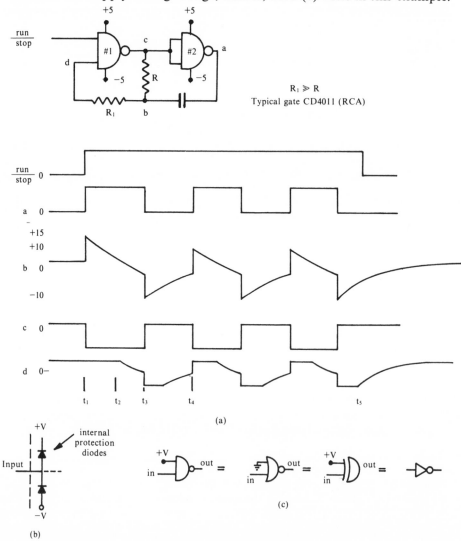

(a)

(b)

(c)

Figure 4-21. Gated multivibrator using CMOS nand gates.

With the run/stop control low (–5volts), "c" will be high (+5 volts) regardless of the "d" input because of the nand function. Gate No. 2 is connected as a simple inverter and with "c" high, "a" will be low and vice versa. In the static condition, "b" and "d" are held high by "c". At t_1 the run/stop control goes high enabling gate No. 1 and, under the stated assumptions, "c" will be low if "d" is greater than the threshold (zero volts) and vice versa. Since "d" is high when the run command is received, "c" will go low and "a" goes high coupling the voltage change at point "a" to point "b" via capacitor C. Since the capacitor had an initial charge corresponding to 10 volts on it, point "b" is driven to +15 volts on this *first* transition as shown in Figure 4-21(a). Resistor R_1 is used to isolate the voltage excursions at "b" that exceed the supply voltages from the diode protection network incorporated at the input of CMOS gates. Generally, R_1 is chosen to be much larger than R so as to not affect the timing. Immediately following the transition at t_1, capacitor C begins to discharge through R because point "c" is low. As the voltage at point "b" comes within the supply voltage range at t_2, the input protection diodes cease to clamp the voltage at point "d" and so point "d" follows point "b." When point "d" crosses zero at t_3, voltages change state as shown in Figure 4-21a and point "b" approaches zero from the negative direction. State changes occur again at t_4, completing the first cycle, and operation continues until the run/stop command goes low.

Under the above assumptions, the frequency of operation is:

$$f = \frac{.72}{RC} \text{ Hz} \qquad\qquad \textbf{(4-12)}$$

Several notes and precautions are applicable:

- 1. If $R_1 \geqslant 10R$, then our assumption that R_2 draws negligible current is justified.

- 2. We have assumed that the threshold voltage is midway between the supply voltages. If this is not the case, different duty cycles will be associated with different thresholds. However, since the duty cycle variations tend to cancel, Equation 4-12 will predict the operating frequency within 5% over the guaranteed range of CMOS gate threshold voltages.

- 3. As shown in Figure 4-21(a), the first timing period t_2–t_3 will be longer than the remaining periods because of the initial charge on capacitor C.

- 4. Any inverter stage could be used for gate No. 2, see Figure 4-21c. If gate No. 2 is replaced with a nor gate, the run/stop control becomes a stop/run control and all the waveforms of Figure 4-21(a) are inverted. For non-gated applications, both gates could be simple inverters and operation would begin upon application of the supply voltages.

- 5. If R_1 is deleted, the circuit will operate, but with several limitations: a) a larger RC time constant is needed for a given frequency; b) the operating frequency becomes more threshold voltage dependent; and c) since the capacitor is clamped by the input protecting diodes, high peak currents will flow during the switching transistion resulting in increased power supply noise.

- 6. When simple inverters are used, both gates must have almost identical threshold voltages to ensure start up. This condition will be met if both gates are on the same chip. If threshold voltages are not matched, gate No. 1 could bias itself to its linear range and since gate No. 2 is not matched, its output is saturated and therefore no gain path exists for the small noise voltages necessary to start oscillations. When the thresholds are matched, if gate No. 1 is in its linear range, then gate No. 2 will also be in its linear range and any disturbance at "c" will be amplified at "a" and fed back to "d" via capacitor C. Since this is positive feedback, latching will occur and multivibrator action will follow. Once started, however, any set of gates will continue to run since only switching operation is necessary.

4.6.2 Oscillator Circuits Using ECL Circuits

Emitter coupled logic gates (ECL), have excellent input and output characteristics and, therefore, oscillator circuits are easily designed. Figure 4-22 shows a multivibrator circuit useful to frequencies in excess of 100MHZ. Circuit operation is identical to that of the circuit of Figure 4-21(a). except that nor gates are used and so the logic polarities are reversed. Because of the low output voltage swing ($\approx$1 volt), resistor R_1 shown in Figure 4-21(a) is not needed in this circuit. Again Equation 4-12 gives the approximate running frequency. Since the input resistance of the 10,000

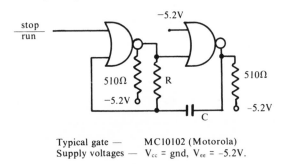

Typical gate —   MC10102 (Motorola)
Supply voltages — V_{cc} = gnd, V_{ee} = –5.2V.

Figure 4-22. Typical gate MC10102 (Motorola) supply voltage V_{cc} = gnd, V_{ee} = –5.2V.

series gates used is 50 kohm, R should not exceed ≈ 10k ohm. Capacitor C should be a silvered mica or NPO ceramic and resistor R should be a carbon composition when operating at the highest frequencies. As before, when simple inverter gates are used, the gates must be on the same chip to ensure startup.

LC Controlled ECL Oscillator

Figure 4-23 shows an LC controlled oscillator. As shown f_{out} is about 35 MHz but the maximum f_{out} of the MC1648 is 225 MHz. LC operation is attractive at high frequencies since it avoids the expense and difficulty of operating crystals in their overtone mode. By proper matching of L and C, very stable frequency operation can be achieved. The $.1\mu f$ capacitors are used to decouple d.c. points in the chip.

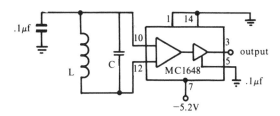

L = 10 turns #26 wire 5/16″ I.D. air core
C = 36 pf silvered mica

Figure 4-23. 35 MHz LC controlled ECL oscillator using the MC1648 (Motorola).

4.6.3   TTL Oscillator

The input and output characteristics of TTL circuits are not as ideal as CMOS or ECL circuits and the simple circuits of Figure 4-21 or 4-22 will not perform well using TTL gates unless very low value resistors are used. The circuit of Figure 4-24(a), however, overcomes most of the limitations by addition of a third gate. Positive feedback occurs between the output of gate No. 3 and the input of gate No. 2 and the overall dc negative feedback loop between the output of gate No. 3 and the input of gate No. 1 ensures that gates No. 2 and No. 3 will be biased into their linear ranges, guaranteeing start up. Unfortunately, this circuit suffers from a very nonsymmetrical duty cycle, typically 20% *off* 80% *on*, because of the clamp diode presently included in all TTL inputs. Thus, capacitor C is charged through the 3k and 1k resistors in series, but it is discharged through the 1k resistor alone.

The circuit of Figure 4-24b achieves symmetric duty cycle by balancing the charge and discharge currents in C by adding the unidirectional current path D_1 and R_2. Typical operating frequency versus capacitance C is tabulated for the circuit of Figure 4-24(b). The circuit of Figure 4-24(a) will have an operating frequency about 2.3 times lower with the same capacitor.

The low power Schottky gates shown have a frequency capability on a par with the standard non-Schottky components, but they have much better input characteristics. For higher frequency operation, a standard Schottky gate could be used, but the resistor values should be reduced by a factor of 10.

The TTL family offers a dual oscillator in low power Schottky (LS 124) and Schottky (S124) versions. These parts feature capacitor or crystal frequency control and can be gated or voltage controlled.

All gates 1/6 LS04   Low power Schottky hex inverter

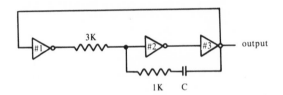

TTL ocillator

Figure 4-24a. TTL oscillator with nonsymmetric output.

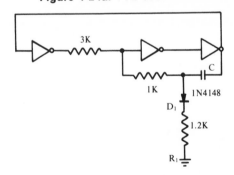

Typical characteristics for circuit of Fig. 4-24b

frequency	capacitance C
285 Hz	1 μf
2857 Hz	.1 μf
263 KHz	1000 pf
2 MHz	100 pf
8.3 MHz	20 pf
10 MHz	10 pf

Figure 4-24b. TTL oscillator with symmetric output.

Chapter 5

Linear Applications

5.1   In this chapter we discuss linear electronic circuits which are made possible through the use of operational amplifiers. We shall describe the operation and the design equations, and the special attention that must be given to certain properties of the operational amplifier which are of primary importance in their effects on the behavior of the final system. The designer must decide what parameters are important in his particular design and he must take them into consideration as was exemplified in Chapter 1, Section 1.7 and in Chapter 4, Section 4.2.1. If the designer is relatively new to the design of electronic systems which utilize I.C.s, he should study Chapter 1 very carefully to ensure that he does not overlook the real-life limitations of operational amplifiers which may have a detrimental effect on his design, unless taken into consideration.

5.2   LINEAR AMPLIFIERS

There are four types of general purpose amplifiers: inverting, noninverting, difference *voltage* amplifiers, and *transconductance* amplifiers. In some applications, an internally compensated amplifier can be used, while in other applications the designer must compensate the operational amplifier externally. Complete analyses and designs with examples and discussions of advantages and shortcomings of the various amplifier types are given in Chapter 1.

5.3   INTEGRATOR

The integrator is used for waveshaping, as for example, to change a square wave to a triangular wave, for signal processing and averaging, and

175

for computation. An integrator circuit is shown in Figure 5-1. So that the reader will be able to design the circuit with the proper selection of components, we shall lead to the circuit of Figure 5-1 step by step, beginning with the ideal amplifier, through the real amplifier and the discussion of special precautions which must be taken to avoid pitfalls.

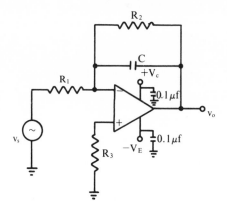

Figure 5-1. Integrator.

5.3.1   The Ideal Integrator

The ideal integrator performs the following mathematical operation:

$$v_o = K \int_0^t v_s dt \qquad\qquad (5\text{-}1)$$

The output v_o is the time integral of the input signal v_s from instant t= 0 as a function of time t. The constant K is a scaling factor which is usually necessary to prevent saturation of the amplifier. If v_s is a dc voltage, then in Equation 5-1 v_s can be taken out of the integral sign, and the integral of dt from zero to t is simply t. Equation 5-1 then becomes

$$v_o = K v_s t$$

which means: if the input is a constant voltage, the output is a ramp function, i.e., the output voltage increases linearly with time (Figure 5-2a). For t = o the output is zero. For t = ∞, the output is infinity (in practice this means that the output reaches a constant value for large t when the amplifier saturates). If the input reverses polarity before the amplifier saturates, the output decreases linearly. By extension it is seen that a square wave input voltage produces a triangular output voltage (Figure 5-2b).

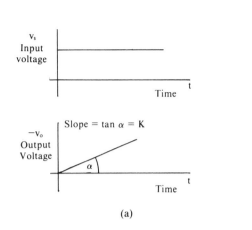

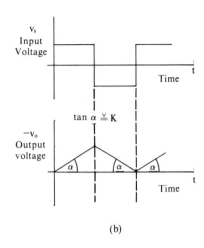

(a)   (b)

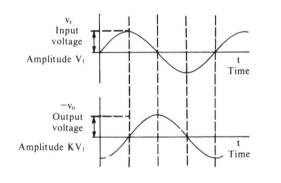

(c)

Figure 5-2. Output voltage of ideal integrator in response to input voltage.

(a) Constant input voltage
(b) Square wave input voltage
(c) Sinewave input voltage

Note that the output $-v_o$ (including a negative sign) is drawn. The input signal is fed into the inverting terminal. A positive input signal causes a negative rate of change of the output.

In terms of frequency, f, the integration operation can be expressed as

$$V_o(f) \;=\; K \;\; \frac{V_s(f)}{j2\pi f}$$

where the input and output voltages are expressed as functions of frequency.

The symbol $\int$ indicates that there is a phase shift between the input and the output voltages. If the input voltage is a sine wave of frequency f, the output voltage is also a sine wave of frequency f lagging the input by 90° (Figure 5-2c).

The ratio of the output voltage to input voltage as a function of frequency is called the *transfer function* of the device. For the integrator, it is

$$\frac{V_o(f)}{V_s(f)} = \frac{V_o}{V_s}(f) = \frac{K}{j2\pi f}$$

The ratio of the amplitudes, disregarding the phase shift is

$$\frac{V_o}{V_s}(f) = \frac{K}{2\pi f}$$

Expressed in decibels the transfer function is

$$\text{Gain (dB)} = 20 \log_{10} \frac{K}{2\pi f}$$

The plot (absolute gain versus frequency) is shown in Figure 5-3. The ideal frequency response has a slope of –20 dB/dec; it has infinite gain for frequence zero; and intersects the 0dB axis at f·k/2π.

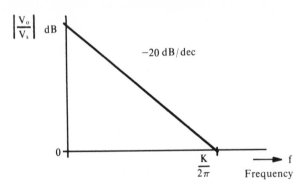

Figure 5-3. Ideal frequency response of integrator.

An integrator is implemented by a circuit shown in Figure 5-4, assuming ideal components. This can be understood as follows. As we have seen in Chapter 1, the gain of the inverting amplifier with a resistor R_1 connected to the inverting input terminal and a feedback resistor R_2 is

$$\frac{V_o}{V_s} = -\frac{R_2}{R_1}$$

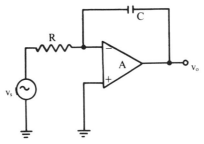

Figure 5-4. Ideal integrator using ideal operational amplifier.

In the integrator, $R_1 = R$ and R_2 is replaced with a capacitor C. The reactance X_c of a capacitor is $1/j2\pi fC$. In analogy with the inverting amplifier we have

$$\frac{V_o}{V_s} = -\frac{1}{R(j2\pi fC)}$$

or, the absolute value of the gain, disregarding sign reversal and phase-shift (i.e., dropping j), is

$$\frac{V_o}{V_s} = \frac{1}{RC2\pi f} \qquad (5\text{-}2)$$

We see that the scaling factor K is equal to $1/RC$.

A physical explanation of the operation of the ideal integrator is obtained by noting that, since the voltage at the inverting input terminal of the ideal amplifier is zero, a current $i=v/R$ flows into the capacitor C charging the capacitor with a charge $q=i\int dt$. The voltage across the capacitor is $v_o=q/C$, or $v_o=(1/C\int)idt$ or $v_o=(1/RC\int)v_s dt$.

5.3.2   The Practical Integrator

The practical integrator is limited by the non-ideal characteristics of the operational amplifier. In particular, special attention must be given to the *finite gain, finite bandwidth, offset voltage, bias current* (or *offset current*), and *output current capability* of the amplifier.

Gain and Bandwidth

Due to the finite gain of the amplifier, the frequency response is truncated for low frequencies beginning with the frequency at which the ideal response intersects the open-loop operational amplifier response as shown in Figure 5-5a. This intersection frequency is given by:

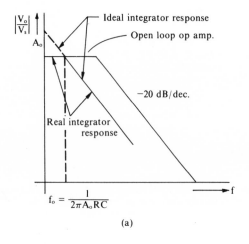

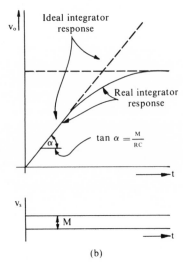

Figure 5-5. Practical integrator response for low frequencies and large values of time.

(a) Frequency response

(b) Time response for step input $v_s = M$.

$$f_o = \frac{1}{2\pi A_o RC}$$

where A_o is the low frequency gain of the op amp. From Figure 5-5a we find that in the low frequency region the frequency response of the integrator is not given by the ideal relation of Equation 5-2, but by:

$$\frac{V_o}{V_s} = \frac{A_o}{1+j2\pi f A_o RC} \tag{5-3}$$

Low frequencies correspond to large values of time. The time response of the integrator for large time values is given by:

$$v_o = v_s A_o (1 - e^{-t/A_o RC}) \tag{5-4}$$

As $t \to \infty$, $v_o \to Av_s$ which indicates open-loop amplifier operation which occurs when the capacitor is fully charged up. Alternately, we can say that since $t \to \infty$ correspondends to $f \to 0$, the capacitor acts as an open circuit.

The slope of the curve expressed by Equation 5-4 is:

$$\frac{dv_o}{dt} = \frac{v_s}{RC}\,e^{I/A_oRC}$$

which for $t \ll A_o RC$ gives:

$$\frac{dv_o}{dt} = \frac{v_s}{RC}$$

from which:

$$v_o = \frac{1}{RC}\cdot v_s dt$$

The time response curve for a step function input, $v_s = M$, is shown in Figure 5-5.

For very short-time values (high frequencies) the finite bandwidth of the amplifier comes into play. For an amplifier open-loop 3dB frequency v_o (corresponding to a time $\tau_o = 1/2\pi f_o$), the time response to a step function of amplitude M is given approximately by:

$$v_o = \frac{M}{RC}(t + \tau_o e - \tau/\tau_o) \qquad\qquad \textbf{(5-5)}$$

and is shown in Figure 5-6. There is a delay of τ_o seconds approached exponentially with a time constant τ_o after the step function has been applied.

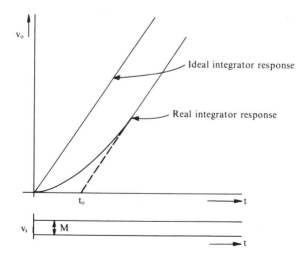

Figure 5-6. Practical integrator response for low values of time.

It is seen that a wide range of operation is available if the circuit is properly designed for the required time and amplitude of integration or frequency and amplitude. The integration must be terminated at a time much less than A_oRC. The maximum permissible time t_m can be computed by a comparison of the exact value and the practical value in terms of the maximum permissible error.

$$\text{error } \% = \frac{AM\,(1 - e^{-t_m/A_oRC}) - t_m/RC}{t_m/RC} \times 100$$

At the other extreme — for small times — the integration approaches the true value only for $t \gg \tau_o$. The quantity τ_o can be determined from the open-loop transfer function of the operational amplifier ($\tau_o = 1/2\pi f_o$, see Figure 5-5a). The error can be computed from a comparison of the correct value $v_o = (M/RC)t$ and Equation 5-5. In the frequency domain, the components must be chosen so that $2f \gg 1\,(2\pi A_oRC)$, where f is the lowest frequency contained in the input signal (see Equation 5-3).

Equation 5-4 has the form of an equation which gives the voltage across the capacitor of a series $R'C'$ network excited by a voltage source. The charging time constant is $T' = R'C'$. Comparing this with Equation 5-4, we conclude that if we connected the component R' and C' to an operational amplifier as in Figure 5-4, the time constant of the system would be given as $T'' = A_oR'C'$ or $T'' = R'C''$ where $C'' = A_oC'$ (making $T'' = A_oT'$). Thus, the operational amplifier gives the capacitor C' an effective capacitance C'' larger than C' by a factor equal to the open-loop low frequency gain of the operational amplifier.

Output Current

The slew rate gives the maximum rate of change of output voltage as limited by the maximum current available to charge the capacitances (usually parasitic capacitances) in the amplifier circuit. The charging current limitation is particularly critical in the integrator because an external capacitor is connected in the circuit which must be charged to provide the necessary output voltage. The maximum possible rate of change of the output voltage is given by:

$$\left(\frac{dv_o}{dt}\right)_{max} = \frac{I_{max}}{C} \qquad (5\text{-}6)$$

where I_{max} is the maximum current that the amplifier can supply. If the input voltage into the amplifier is a sinusoidal voltage $v_s = V_m \sin 2\pi ft$, then the output voltage of the integrator is $v_o = -(V_m/RC2\pi f) \cos 2\pi f$ and the maximum rate of change of the output is obtained from

$$\frac{dv_o}{dt} \quad \frac{V_m}{RC}\sin 2\pi ft \text{ for } \pi ft = \frac{\pi}{2} \text{ as } (\frac{dv_o}{dt_{max}}) = \frac{V_m}{RC} \cdot$$

Comparing this result with Equation 5-6, we conclude that to allow an output whose accuracy is not limited by the output capability of the amplifier we must make certain that:

$$\frac{V_m}{R} \leq I_{max}$$

This condition puts a lower limit on R for a given amplitude V_m of the input voltage.

Another lower limit on R is imposed by the maximum leading allowed on the signal source. The higher of the two values imposed by these limits must be taken for the lowest permissible value of R. It is interesting to note that, unlike the voltage amplifier case, where the product $V_m\omega$ must be limited, no limit is put on the frequency as far as the available current is concerned, but only on the peak voltage. The reason is that in the integrator circuit, while the rate of change of the output increases with frequency, the amplitude of the output decreases by the same factor.

Offset Voltage

The offset voltage v_{offset} appears at the output directly, as well as in the form of a ramp function as it charges the capacitor according to $(1/RC\int)v_{offset}dt$. This output appears as an error and limits the accuracy of the integrator. If no precautions are taken, then when the amplifier is idle between signal applications, the offset voltage will cause an increasing output until the amplifier saturates, making it useless. To avoid this situation, a resistor (R_2 in Figure 5-1) must be connected across the capacitor. At low frequency the reactance of the capacitor is large in comparison with the resistor, and the integrator acts as an amplifier with gain R_2/R_1 which keeps the output in response to the offset voltage relatively small. At high frequencies the resistor is shunted by the smaller capacitive reactance and the circuit acts as an integrator. The transition from amplifier to integrator can be defined as occurring at the frequency f_c,

$$f_c = \frac{1}{2\pi R_2 C}$$

where the absolute value of the capacitive reactance equals the resistance

$$R_2 = \frac{1}{2\pi f_c C}$$

The minimum size of R_2 is determined by the required accuracy for a given frequency. For a frequency $f = 10f_o$ for an integrator with a d-c gain of 10, the accuracy is better than 1%.

Another way to look at this is to say that the capacitor must have an opportunity to discharge, but the discharge circuit R_2C must have a time constant at least ten times longer than the integrator time constant R_1C.

Bias or Offset Current

As we have seen in Chapter 1, the bias current multiplied by the resistance through which it flows produces a voltage at the operational amplifier terminals with an effect similar to the offset voltage effect. To minimize the error caused by the bias current, a resistor $R_3 = R_1 \| R_2$ must be connected to the noninverting terminal as shown in Figure 5-1. The error in the output due to bias currents is then only a function of the offset current. If $R_2 \geqslant R_1$, then for simplicity we assume for the following demonstration that $R_1 \simeq R_1 \| R_2$ so that we can assume a voltage $R_1 I_{offset}$ caused by the offset current. This will cause an output $(1/R_1C)\int^t R I_{offset} dt$ or $1/C\int^t_{offset} dt$. The effect of the offset current can be reduced by increasing C while decreasing R_1 to keep the product R_1C constant. The minimum value of R_1 was discussed in connection with Equation 5-6 and the related paragraph.

5.3.3   Pitfalls to Avoid

In addition to the considerations given to the effects of the characteristics of real amplifiers as described in the previous section, the designer's choice of an amplifier must be guided also by less obvious characteristics: the common-mode and differential-mode input voltage ranges. Because of the capacitive coupling of the output to the input of the amplifier, transients occurring at the output are coupled back to the input and either common-mode or differential-mode voltage limits can be exceeded. Furthermore, if fast rising or falling signals are applied to the amplifier, such as a square wave, then, momentarily, high voltages, exceeding safe limits, can appear at the inverting input, since the integrator cannot respond instantaneously. The maximum common-mode and differential-mode input specifications of the amplifier must, therefore, be considered carefully.

Depending on the required accuracy, it may be necessary to use FET input stage amplifiers because of the low bias current requirements. Provisions should always be made for d-c feedback such as provided by R_2 in Figure 5-1, to prevent increasing outputs resulting from drifts and offset voltages. For long period integration, capacitors must be selected with low dielectric leakage current. Polystyrene and Teflon capacitors are suitable.

For high-speed integration small values of capacitors are needed and Mylar or silver-mica capacitors can be used.

5.3.4 Numerical Example

Reasonable component values for an integrator configuration of Figure 5-1 designed to operate at 1 kHz are: $R_1 = 39$ K, $R_2 = 390$K, $R_3 = 39$K, $C = 0.03$ μF. The 0.1 μF capacitors are connected for decoupling from the power supplies. The d-c gain is 20 dB and becomes ineffective for frequencies above 10 or 15 Hz.

5.4 DIFFERENTIATOR

A differentiator circuit is shown in Figure 5-7. To show the designer how he can arrive at a design to meet his particular requirements, we shall begin with an ideal differentiator and progress to a practical differentiator considering practical design limitations.

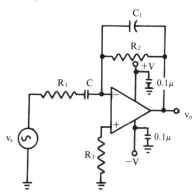

Figure 5-7. Practical differentiator.

5.4.1 The Ideal Differentiator

The differentiation operation (multiplied by a constant scaling factor K) is written

$$v_o = K \frac{dv_s}{dt}$$

The output voltage v_o is proportional to the time derivative of the input voltage v_s. In terms of frequency f the relationship is

$$V_o = jK2\pi fV_s$$

The symbol j indicates a phase shift between input and output, the output leading the input by 90°. When only magnitudes are of interest the j is dropped from the equation. The gain expression in terms of magnitudes is

$$\frac{V_o}{V_s} = K2\pi f$$

The gain is proportional to the frequency of the input voltage (Figure 5-8).

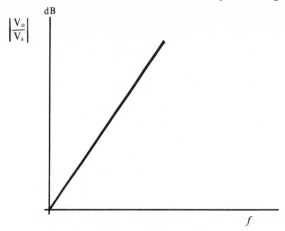

Figure 5-8. Gain-frequency response of ideal differentiator.

In theory, assuming an ideal amplifier with infinite gain and infinite bandwidth, such a gain is displayed by a system shown in Figure 5-9. If the capacitor were replaced with a resistor, the gain of the amplifier would be given by the negative of the resistance ratio (feedback resistance divided by the resistance between the signal source and the inverting input). The reactance of capacitor C is $X_c = 1/j2\pi fC$. In analogy to the gain of an inverting amplifier the gain of the differentiator is $(V_o/V_s) = -j2\pi fCR$. The gain ratio of the magnitudes, disregarding the phase shift, is

$$\frac{V_o}{V_s} = -RC2\pi f$$

We note that the proportionality constant is $K = -RC$. The negative sign is associated with the inverting input terminal of the operational amplifier.

In practice, the implementation of such a gain function is impossible as any operational amplifier attenuates higher frequencies. Furthermore, a

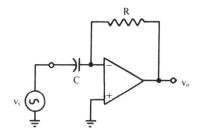

Figure 5-9. Ideal differentiator.

purely capacitive input as shown in Figure 5-9 is undesirable for two reasons: (a) higher frequencies are amplified more than lower frequencies. Noise whose frequency content may be well above the signal frequencies is amplified excessively rendering the system useless; (b) the purely capacitive input is usually an unsuitable load for the signal source.

5.4.2   The Practical Differentiator

To alleviate the problems associated with the input capacitor described in the previous section, one adds a resistor R_1 as shown in Figure 5-7. In practice, one is interested to differentiate signals up to a certain frequency f_1; the linear relationship of Figure 5-8 is therefore required only up to this frequency. The value of R_1 is chosen so that together with C a break frequency is obtained at f_1. The transfer function of the circuit of Figure 5-7, ignoring C_f for the time being, is:

$$\frac{V_o}{V_s} = - \frac{R}{R_1 + \dfrac{1}{j2\pi fC}} = - \frac{j2\pi fRC}{1 + j2\pi fR_1C} \qquad (5\text{-}7)$$

Thus, we introduced a breakpoint at $f_1 = 1/2\pi R_1C$,. and R_1 is chosen accordingly. The response curve of Equation 5-7 is shown in Figure 5-10 together with the open loop response curve of the amplifier. This curve shows that d-c signals are blocked, low frequency signals up to $f_1 = 1/2\pi R_1C$ are differentiated, signals of frequencies between f_1 and f_b are passed, and signals of frequencies greater than f_b are attenuated. The frequency f_b is the intersection frequency of the open-loop and the closed-loop frequency response curves. If the intersection occurs at a steeper than –20 dB/dec slope of the open-loop response curve, the amplifier must be compensated as described in Chapter 1.

There is no good reason to pass signals beyond the frequencies which

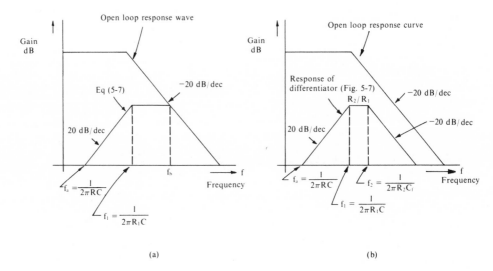

(a)   (b)

Figure 5-10. Gain-frequency response curve of a practical differentiator.

(a) Without feedback capacitor Cf
(b) With feedback capacitor Cf to reduce high frequency noise.

need be differentiated. Passing higher frequencies makes the differentiator prone to noise and oscillations. It is therefore good practice to add the capacitor C_f as shown in Figure 5-7 to introduce a second break frequency at

$$f_2 = \frac{1}{2\pi RC_f}$$

The output magnitude is limited to the input magnitude multiplied by R_2/R_1. This is the value of the flat region between f_1 and f_2 in Figure 5-10(b).

5.4.3   Pitfalls to Avoid

Although the input impedance of the circuit in Figure 5-7 is not purely reactive, it could have a large reactive component depending on the relative values of R_1 and $2\pi fC$ where f is the frequency of the signal which must be differentiated. If the signal source requires a purely resistive load, a more expensive circuit using two operational amplifiers must be used as described in Section 5.5.

The resistor R_1 may have a small value of a few tens ohms, depending on the frequencies involved, and the output resistance of the signal source cannot always be neglected. It is quite possible to have in effect a circuit as shown in Figure 5-7 without adding an external resistor R_1 due to this output resistance. It is important to note that R_1 does not provide a dc path for bias current because C is in series with it. Hence R_3 must be chosen equal to R to equalize the bias current effect.

The slew rate of the amplifier must be considered in conjunction with the frequency and amplitude of the required output. If the input is a triangular wave, the output should be a rectangular wave. The steepness of the rising and falling edges is limited by the slew rate of the amplifier.

5.4.4   Numerical Example

For a signal of 1kHz the circuit of Figure 5-7 is suitable with the following components: R = 10KΩ, C = 0.1 μF, R_1 = 470Ω, R_2 = 10 KΩ, C_f = 3300 pF. The following frequencies (Figure 5-10b) are then obtained: f_a = 159 Hz; f_1 = 3.386 kHz, f_2 = 4.823 kHz. The maximum output magnitude is equal to the input magnitude multiplied by 21.277.

5.5   TWO OPERATIONAL AMPLIFIER—RESISTIVE INPUT DIFFERENTIATORS

Figure 5-11 shows a differentiator which has a resistive input impedance. Since points a and a′ are at virtual ground potential, the current drawn from the signal source v_s is limited only by R_1 and R_2. Hence, the input resistance of the differentiator is given by:

$$R_{in} = R_1 \| R_2 = \frac{R_1 R_2}{R_1 + R_2}$$

the transfer function of the differentiator is

$$\frac{V_o}{V_s} = - \frac{R_3}{R_2} \frac{j2\pi f R_1 C}{(1 + j2\pi f R_1 C)(1 + j2\pi f R_3 C_f)}$$

The gain-frequency response curve has the same form as this shown in Figure 5-10b with the following significant frequencies:

$$f_a = 1/(2\pi \frac{R_1 R_3}{R_2} C); \quad f_1 = 1/(2\pi R_1 C); \quad \text{and } f_2 = 1/(2\pi R_3 C_f).$$

The scaling factor is $$K = - \frac{R_3}{R_2} R_1 C$$

The circuit, therefore, performs differentiation, and the break frequencies are chosen according to the signal frequencies and are implemented by a choice of R's and C's according to the equations for the break frequencies.

5.5.1 Numerical Example

For a signal of 5 kHz the circuit of Figure 5-11 is suitable with the following components: $R_1 = 2.7\,K\Omega$, $R_2 = 180\,K\Omega$, $R_3 = 240\,K\Omega$, $R_4 = 1.3\,K\Omega$, $R_5 = 100\,k\Omega$, $C = 0.01\,\mu F$, $C_f = 100\,pF$. The resulting significant frequencies (Figure 5-10b) are: $f_a = 3.9$ kHz, $f_1 = 5.9$ kHz, $f_2 = 6.6$ kHz. The component values were obtained by first choosing the desired frequencies, computing the required component values, rounding off the component values to standard values, and finally recomputing the frequencies with the standard component values.

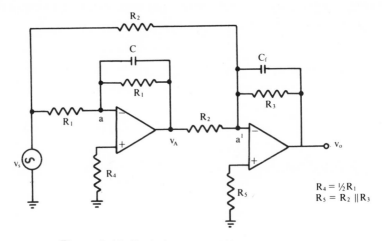

Figure 5-11. Resistive input differentiator.

5.6 INSTRUMENTATION AMPLIFIER

The amplifier in Section 1.6.4 offers good common mode rejection and is relatively inexpensive. It suffers from the disadvantage of relatively low input impedance associated with the inverting output. For symmetrical amplification of the differential input, the circuitry associated with the noninverting input results in the same low input impedance.

The circuit of Figure 5-12 alleviates the low input impedance circuitry by preceding the differential amplifier stage with another stage designed specifically to offer high input impedances. This is achieved by using two amplifiers and connecting the differential input signal to the noninverting input terminals of both amplifiers. The output of these

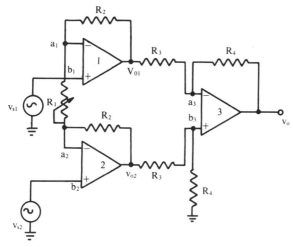

Figure 5-12. Instrumentation amplifier.

amplifiers is connected to the differential stage. The impedance between the two input terminals is

$$\frac{2R_iA}{1 + \dfrac{R_2}{2R_1}}$$

where R_i and A are the input resistance and the gain of the open-loop operational amplifiers.

Differential Gain ($v_{s1} = -v_{s2}$)

The gain of the amplifier is equal to the product of the gains of the two stages. The second stage (operational amplifier 3) is connected in the inverting configuration and its gain is

$$K_2 = -\frac{R_4}{R_3}$$

The amplifiers of the first stage (op. amps. 1 and 2) are connected in the noninverting configuration. The gain of this stage is

$$K_1 = (1 + \frac{R_2}{R_1/2})$$

The resistance $R_1/2$ replaces R_1 in the conventional noninverting amplifier gain equation. In the conventional amplifier, R_1 is connected to ground. This is not the case here. But the potential midway between the ends of R_1 is zero (or in effect at ground potential). The reasoning leading to this statement follows.

Since essentially $v_{a1} = v_{b1}$ and $v_{a2} = v_{b2}$, the voltages at the terminals of R_1 are v_{s1} and $v_{s2} = -v_{s1}$. For $v_{s1} = 0.2$ V, for example, the voltage terminal at terminal a_1 of $R_{1\pi}$ is 0.2 V and at terminal a_2 is –0.2V. There will be a current flowing through R_1 from a_1 to a_2 causing a uniform voltage drop along R_1, resulting in a voltage of zero halfway down the resistor. The voltage gain associated with operational amplifier 1 is therefore

$$\frac{v_{o1}}{v_{s1}} = (1 + \frac{R_2}{1/2\ R_1})$$

and that of operational amplifier 2 is

$$\frac{v_{o2}}{v_{s2}} = (1 + \frac{R_2}{1/2\ R_1})$$

Combining these two equations gives

$$v_{o1} - v_{o2} = (1 + \frac{2R_2}{R_1})\ (v_{s1} - v_{s2})$$

Combining this equation with that of the second stage

$$v_o = -\frac{R_4}{R_3}\ (v_{o1} - v_{o2})$$

we have for the output of the overall amplifier

$$v_o = -\frac{R_4}{R_3}\ (1 + \frac{2R_2}{R_1})\ (v_{s1} - v_{s2})$$

Common-Mode Gain $(v_{s1} = v_2)$

To study the common-mode gain, we let $v_{s1} = v_{s2}$. We then have

$$v_{b1} = v_{a1} = v_{b2} = v_{a2}$$

No current flows through R_1 and therefore no current through the R_2's. Consequently,

$$v_{o1} = v_{s1}$$

and

$$v_{o2} = v_{s2}$$

This shows that the common-mode gain of the first stage of the amplifier is unity. The second stage has a common-mode gain determined by the CMRR as was discussed in Chapter 1 and the overall common-mode rejection of the amplifier is determined by the second stage alone.

5.7 BRIDGE AMPLIFIERS

In instrumentation circuits where high sensitivity is needed, bridge circuits often are employed. Examples are strain and temperature measurement circuits using strain gauges or thermistors. The strain or temperature is measured in terms of a change in the resistance of the transducer. Either one transducer or two transducers can be used in one or two branches of the bridges, respectively. A single transducer arrangement is shown in Figure 5-13. The output voltage v_s of the bridge is an indication of the measured quantity.

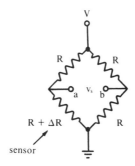

Figure 5-13. Single transducer bridge arrangement.

The output of the bridge can be connected to the input terminals of an instrumentation amplifier like the one shown in Figure 5-12 so that $v_s = v_{s1} - v_{s2}$. As we shall show in the next paragraph, the output of the amplifier is proportional to $\Delta R/R$ (which in turn is proportional to the monitored quantity, such as temperature) provided $\Delta R \ll R$.

Since the input impedance of the instrumentation amplifier is practically infinite, the output of the bridge is essentially open-circuited. The output voltage can therefore be obtained using voltage division relationships: at terminal "a", $v_a = [(R + \Delta R)/(2R + \Delta R)]V$, at terminal "b", $v_b = (R/2R)V$. Hence, we have for $v_s = v_a - v_b$,

$$v_s = (\frac{R + \Delta R}{2R + \Delta R} - \frac{R}{2R}) \ V = \frac{\Delta R}{2(2R+\Delta R)}$$

or

$$v_s = \frac{\Delta R}{4R} \ \text{for } R \gg \Delta R$$

Since the output of the amplifier (Figure 5-12) is proportional to the differential input, the output voltage v_o is proportional to ΔR for $R \gg \Delta R$,

$$v_s = K\Delta R$$

where

$$K = \frac{R_4}{4RR_3}\left(1 + \frac{2R_2}{R_1}\right)$$

The resistance values of the bridge must be well-matched for the equation to be correct. Drawbacks of this measurement system are that (a) it is expensive since it utilizes three operational amplifiers with the associated circuitry, and (b) it is linear only for small variations in the sensor resistance.

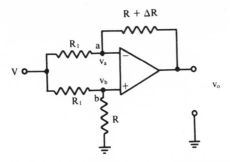

Figure 5-14. Wide-deviation bridge ampilifier.

Both these drawbacks are overcome with the circuit shown in Figure 5-14. The bridge is a modified version of the original bridge of Figure 5-13, in that $R + \Delta R$ and R do not have a common point. The output of this amplifier is derived by equating the current entering node "a" to the current leaving the node. Since current between two points is equal to the difference in voltage between the two points divided by the resistance between the two points, we have

$$\frac{V - V_a}{R_1} = \frac{V_a - V_o}{R + \Delta R}$$

The voltage at node "b" is by the voltage dividing network at the node

$$V_b = V\frac{R}{R_1 + R}$$

Noting that $v_a = v_b$, we solve for v_o from the first equation and substitute the second equation for v_a,

$$V_o = \frac{\Delta R}{R_1 + R}V$$

Thus, the output is directly proportional to ΔR which is proportional to the monitored quantity. Another way to write this equation is

$$v_o \;=\; \frac{R\;\delta R}{R_1 + R}\;V$$

where $\delta = \Delta R / R$ gives the relative change in the resistance of the sensor. Since the output of this amplifier is proportional to ΔR for large as well as small deviations (no restriction of $\Delta R / R \ll 1$ is required) this amplifier configuration can be used for highly sensitive transducers such as semiconductors or certain thin-film strain gauges.

5.8   CURRENT-TO-VOLTAGE CONVERTER

The operational amplifier is suitable to measure current with little error by injecting the current directly into the inverting terminal. The appropriate circuit is shown in Figure 5-15a. The basic circuit (Figure 5-15b)

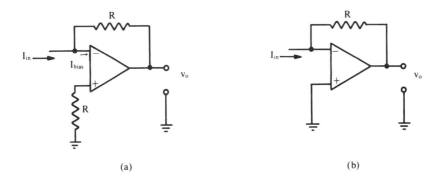

(a)   (b)

Figure 5-15. Current-to-voltage converter.

(a) Circuit compensated for minimum bias current error.
(b) Basic circuit configuration.

shows that since the inverting input terminal is at virtual ground, the current I_{in} to be measured is not impeded by the measuring circuit. Under ideal conditions (no bias current and infinite input impedance at the inverting terminal), the output v_o adjusts itself so that the current in R is equal in magnitude to I_{in}:

$$v_o \;=\; -R I_{in}$$

and the output voltage is directly proportional to the current to be measured. The value of R is chosen so that the expected current produces the desired output voltage. If, for example, a voltage of 200 mV is desired with the current of 20 μA, then

$$R = \frac{200 \times 10^{-3}}{20 \times 10^{-6}} = 10 \text{ K ohm}$$

Since, in fact, the terminal does draw bias current, I_{bias}, the nodal equation gives

$$v_0 = -RI_{in} + RI_{bias}$$

The error introduced by the bias current is minimized by adding a resistor to the noninverting terminal as shown in Figure 5-15a. Here the output voltage is given by

$$v_0 = -RI_{in} + RI_{offset}$$

where I_{offset} is the bias offset current which is considerably smaller than the bias current. The amplifier must be chosen so that the offset current is much smaller than the current to be measured to obtain the desired accuracy. For the measurement of nanoamps and picoamps it is usually necessary to use an operational amplifier with an FET input stage.

Pitfalls to Avoid

Since the operational amplifier is usually thought of as a voltage amplifier, there is a tendency by designers first to convert the current to be

DO NOT USE THESE CIRCUITS!

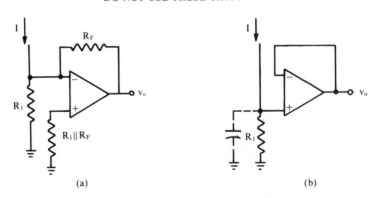

(a)   (b)

Figure 5-16. Circuits for the measurement of current to be avoided.

(a) Inverting amplifier (b) Voltage follower.

measured into a voltage and then to amplify the voltage, using one of the circuits shown in Figure 5-16. Such circuits have serious disadvantages and should be avoided. Resistor R_1 introduces an impedance and changes the current under measurement. Furthermore, the offset voltage will be amplified and introduce additional error. An additional disadvantage is the slow response time caused by the stray capacitance shown in dotted lines.[1] Particularly when the current to be measured is very small, R_1 must be very large. As I varies, the capacitor must be charged and discharged giving a slow response to the amplifier. This limitation does not exist in the circuit of Figure 5-15 since the inverting terminal is at virtual ground voltage and the stray capacitance does not need to be charged or discharged with changes in measured current.

5.9 PHOTOSENSOR AMPLIFIERS

Photodiodes can be used in two different modes to measure light intensity: in the reverse-biased mode and in the short-circuited mode. The I-V characteristic curves of a photodiode are shown in Figure 5-17. It can be

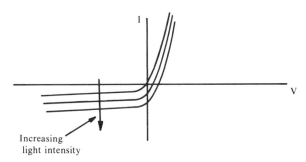

Figure 5-17. Photodiode characteristic curves (not to scale).

seen that in the reserve-biased mode the diode approaches a current generator, i.e., the current is nearly independent of applied voltages. Since, however, the current is slightly dependent on the applied reverse voltage, the current as a function of light intensity is also slightly dependent on the applied reverse voltage. The circuit configuration of Figure 5-15a should be used when a photodiode is used as the sensor. Photodiode amplifiers are shown in Figures 5-18a and 5-18b. A photoconductive cell transducer with amplifier is shown in Figure 5-18c. Note that in all cases the voltage across the light sensitive device is constant so that the current through the device depends only on the light. The frequency response is not limited by stray capacitance at the terminals as was discussed in Section 5.8 in connection with Figures 5-15 and 5-16.

---

[1] This stray capacitance includes any output capacitance of the current source, for example, the junction capacitance of a reversed biased photodiode.

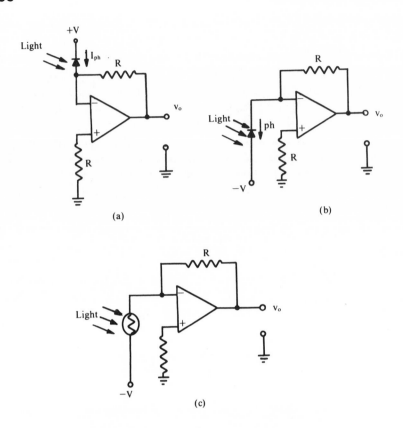

Figure 5-18. Light detector amplifier.
(a) Photodiode returned to positive voltage.
(b) Photodiode returned to negative voltega.
(c) Photoconductive cell.

Pitfalls to Avoid

Because of misuses that have been made by some designers in the past, we emphasize that circuits of the configurations shown in Figure 5-16 should not be used in conjunction with photoconductive diodes. In addition to the disadvantages of such an arrangement discussed in Section 5.8, when a photodiode is used in the branches marked I in Figure 5-16, the voltage across the diode varies as a result of the current dependent voltage across R_1.

Instead of photodiodes, phototransistors can be used with the collector-emitter terminals replacing the diode terminals. The base is biased at a voltage roughly midway between V or (–V) and ground, and the light modulates the collector-emitter current. Such an arrangement adds current gain, but is not a preferred arrangement since the characteristics of transistors vary widely from one unit to another (same type number) while those of diodes do not.

In all the circuits of Figure 5-18 the output voltage is proportional to the photosensitive device current (neglecting bias current), the sign depends on the particular configuration. Considering the bias current, the output-input relation is

$$v_o \;=\; \pm RI + RI_{offset}$$

The diode current is a combination of the photocurrent and the temperature sensitive leakage current. If the sensor is to be used over a wide range of temperature, the leakage current can introduce considerable error. This problem is alleviated if the diode is used in the *short-circuit mode*. With zero voltage across the diode there is no leakage current and the short-circuit current is equal to the photocurrent (Figure 5-17). A circuit which utilizes the photodiode in the short-circuit mode is shown in Figure 5-19. In reality the amplifier offset voltage appears across the diode. Since this voltage is very small, the leakage current is at least two orders of magnitude smaller than in the reverse bias mode. If the offset voltage is neglected, the output voltage of the amplifier is

$$v_o \;=\; 2RI_{ph}$$

where I_{ph} is the diode photocurrent. If, for example, R = 5 MΩ, the conversion factor is $10V/\mu A$.

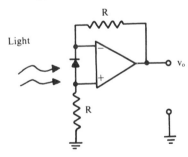

Figure 5-19. Photodiode sensor amplifier for operation in the short-circuit mode.

5.10   THE VOLTAGE FOLLOWER

A voltage follower circuit is shown in Figure 5-20a. The output voltage is equal to the input voltage. The amplifier is used for isolation or impedance transformation. It is a special case of the noninverting amplifier with the feedback path short-circuited and with no path between the inverting terminal and ground. The input resistance is equal approximately to the open-loop input resistance of the operational amplifier times the open-loop gain, and the output resistance is equal approximately to the open-loop output resistance of the operational amplifier divided by the open-loop gain. The voltage follower clearly is useful to interface a high output impedance signal source to a low impedance load. To minimize the offset imbalance resulting from bias currents, considering the output impedance of the signal source, a resistor is sometimes connected to the noninverting terminal as shown in Figure 5-20b.

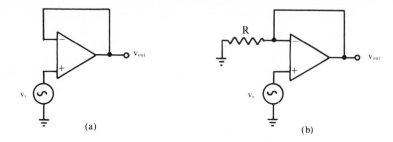

(a)   (b)

Figure 5-20. The voltage follower.
(a) Simplest and least expensive
(b) With resistor to reduce imbalance.

5.11   FOUR-QUADRANT MULTIPLIER

The operational transconductance amplifier makes the construction of a four-quadrant multiplier relatively simple. The principle of operation can be explained with the help of Figure 5-21. All three amplifiers are transconductance amplifiers. Because of the high output impedances, the amplifiers perform as controlled current sources. The output voltage across R_L is therefore

$$v_o = \left[i_{o(1)} + i_{o(2)} \right] R_L$$

Amplifier "1" is used in the inverting mode for which

$$i_{o(1)} = -g_{(1)}v_x$$

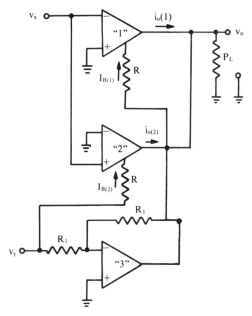

Figure 5-21. Basic configuration for a four-quadrant multiplier. (Reprinted by permission of RCA Solid State Division)

and amplifier "2" is used in the noninverting mode for which

$$i_{o(2)} = g_{(2)}v_x$$

giving for the output

$$v_o = \left[-g_{(1)}(v_x) + g_{(2)}(v_x) \right] = v_x \left[g_{(2)} - g_{(1)} \right]$$

The transconductance of the amplifiers is controlled by the bias currents:

$$i_{B(1)} \simeq \frac{(V^-) - v_y}{R}$$

(note that amplifier "3" is connected as a unity gain converter)

$$I_{B(2)} \simeq \frac{(V^-) + v_y}{R}$$

The transconductances can therefore be written

$$g_{(1)} \simeq K \left[(V^-) - v_y \right]$$
$$g_{(2)} \simeq K \left[(V^-) + v_y \right]$$

where K is a proportionality constant. This gives for the output voltage,

$$v_o \simeq 2KR_L v_x v_y$$

A complete circuit including provisions for the necessary adjustments is shown in Figure 5-22. The adjustment procedure is as follows: with both v_x and v_y set to zero, terminal 10 is connected to terminal 8. This disables amplifier "2" and permits adjustment of the offset voltage of amplifier "1" to zero by means of the potentiometer R_c. The short between terminals 10 and 8 is then removed, and terminal 15 is connected to terminal 8. This disables amplifier "1" and amplifier "2" is zeroed by means of potentiometer R_D. Finally, a-c signals are applied to terminals X and Y and potentiometers R_A and R_B are adjusted for symmetrical output signals.

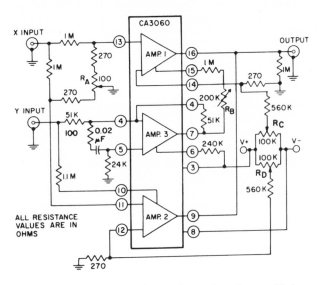

All resistance values in ohms unless otherwise specified.

Figure 5-22. A CA3060 four-quadrant multiplier circuit.
(Reprinted by permission of RCA Solid State Division)

5.12   A-C COUPLED AMPLIFIERS

At times it is necessary to remove the d-c component from the previous stage or signal source before the signal is processed further. This is done by a-c coupling.

5.12.1   Inverting Amplifier

The circuit is shown in Figure 5-23a. The configuration is the same as that of the d-c inverting amplifier, except that a capacitor is inserted to block the d-c voltage. The size of the capacitor is determined by the lowest frequency which must be amplified. The gain of an inverting amplifier is in general terms of impedance Z_1 and Z_2 (Figure 5-23b),.

$$\frac{V_o}{V_s} = -\frac{Z_2}{Z_1}$$

In the circuit of Figure 5-23a, $Z_2 = R_2$ and $Z_1 = R_1 + 1/j2\pi fC_1$. The transfer function is therefore

$$\frac{V_o}{V_s}(j2\pi f) = -\frac{j2\pi fR_2C_1}{1+j2\pi fR_1C_1}$$

It is seen that for f=0, $(V_o/V_s)=0$; for $f\to\infty$, $(V_o/V_s)\to -R_2/R_1$; the 3 dB break frequency is $f_{3db} = 1/2\pi R_1C_1$. For high frequencies $(f > f_{3db})$ the capacitor is essentially a short circuit and the input and output impedances are identical to those of the d-c coupled inverting amplifier. The input impedance for example, is therefore equal to R_1.

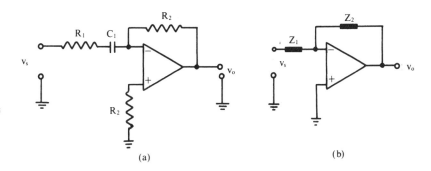

Figure 5-23. (a) A-C coupled inverting amplifier.
(b) Inverting amplifier with impedance networks.

There will be a d-c offset output voltage caused by the combination of the d-c input offset voltage and the input offset current. Because of the presence of C_1, the d-c voltage between the input terminals will appear at the output with a gain of unity. Thus, the offset voltage at the output is

$$V_{o(offset)} = V_{i(offset)} + R_2 I_{(offset)}$$

Note that the resistor connected to the noninverting terminal to minimize the offset caused by the bias currents is equal to the feedback resistor R_2; not to the parallel combination $R_2 \| R_1$ as would be the case for a d-c coupled amplifier, since the capacitor blocks the bias current which is therefore supplied to the inverting terminal only through the feedback resistor.

5.12.2   Noninverting Amplifier

An a-c noninverting amplifier is shown in Figure 5-24. The gain of the amplifier as function of frequency is

$$\frac{V_o}{V_s}(j2\pi f) = \frac{1}{1+\dfrac{1}{j2\pi fR_3C_1}} \times \frac{1+j2\pi f(R_2+R_3)C_2}{1+j2\pi fR_2C_2}$$

The capacitor C_2 can be replaced with a short circuit in which case the compensation resistor R_1 at the noninverting terminal is replaced with the parallel combination $R_1 = R_2 \| R_3$ to minimize bias effects. The gain equation is then

$$\frac{V_o}{V_s}(j2\pi f) = \frac{1}{1+1/j2\pi fR_1C_1}\left(1+\frac{R_3}{R_2}\right)$$

For high frequencies $(f > 1/2\pi R_1C_1)$ the transfer function reduces to that of the d-c coupled noninverting amplifier.

$$\frac{V_o}{V_s} = \left(1+\frac{R_3}{R_2}\right)$$

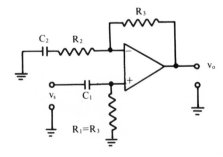

Figure 5-24. A-C coupled noninverting amplifier.

Note that it is necessary to provide d-c paths to both input terminals for bias currents. It is therefore not possible to remove R_1 from the circuit, which would make the circuit analogous to the d-c coupled noninverting amplifier with the associated high input impedance. With R_1 present, the input impedance is limited to R_1. A high input impedance, noninverting amplifier is presented in the next section.

5.12.3 High Input Impedance, A-C Coupled, Noninverting Amplifier

A "bootstrapped" high input impedance, a-c coupled amplifier is shown in Figure 5-25. Bias current to the inverting terminal is supplied

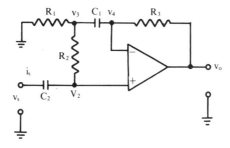

Figure 5-25. High input impedance A-C coupled amplifier.

through R_3, and to the noninverting terminal through R_1 in series with R_2. At low frequencies C_2 provides high impedance. It is interesting to see what happens at higher frequencies $(f>1/2\pi(R_1+R_2)C_2)$ where C_2 provides essentially a short circuit. With C_1 of the same order of magnitude as C_2 or larger, at higher frequencies the voltage drop across C_1 approaches zero, causing R_2 to be effectively connected between the input terminals of the amplifier. The voltage across R_2 is then only the very small offset voltage which implies a very small current through R_2, and therefore a high input resistance.

$$R_i = v_s/i_s$$

As usual, this is true, provided the open-loop gain of the amplifier is very high causing $v_2 \simeq v_4$. At very high frequencies, for which the open-loop gain of the amplifier becomes small, the voltage across the amplifier input terminals increases and substantial currents can flow through R_2 causing a reduced input impedance.

For high frequencies, for which the capacitors offer essentially no reactance,

$$v_o/v_s = 1 + \frac{R_3}{R_1}$$

The input impedance is

$$Z_{in} = \frac{1}{j2\pi fC_2} + R_1 + R_2 + j2\pi fC_1R_1R_2$$

At low frequencies, the input impedance is high because of C_2 and at high frequencies because of C_1 except, as noted earlier, the input impedance decreases again at frequencies for which the open-loop gain of the amplifier decreases.

5.12.4 Differential Instrumentation Amplifiers

Figures 5-26 and 5-27 show, respectively, an a-c differential amplifier and an a-c high input resistance instrumentation amplifier. The output voltage of the differential amplifier, Figure 5-26, is

$$v_o = \frac{j2\pi fR_2C_1}{1 + j2\pi fR_1C_1} (v_{s2} - v_{s1})$$

At high frequencies the equation reduces to $v_o = (R_2/R_1)(v_{s2} - v_{s1})$.

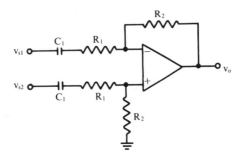

Figure 5-26. A-C coupled differential amplifier.

The output of the instrumentation amplifier (Figure 5-27) is given by

$$v_o = (1 + \frac{2R_2}{R_1})(\frac{j2\pi fR_4C_3}{1 + j2\pi fR_3C_3}) (V_{s2} - V_{s1})$$

The d-c decoupling is done between the first and the second stages so that the signal source is not loaded by a capacitor. At high frequencies, the output equation reduces to

$$V_o = \frac{R_4}{R_3}\left(1+\frac{2R_2}{R_1}\right)(v_{s2}-v_{s1})$$

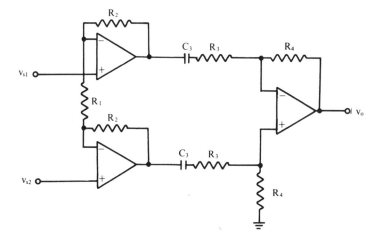

Figure 5-27. A-C instrumentation amplifier.

Chapter 6

Nonlinear Applications

6.1 This chapter is divided into two parts. Part No. 1 discusses using operational amplifiers in conjunction with nonlinear devices and passive elements to generate special transfer functions. Part No. 2 deals with voltage comparators and some typical applications for these devices.

6.2 NONLINEAR CIRCUITS

6.2.1 Precision Voltage Rectification

The simple half-wave rectifier shown in Figure 6-1(a) has very limited application in low-level circuits because of the diode voltage drop during forward conduction. About .7 volts (silicon diode) are required at the input before any output voltage is noticed. Furthermore, this voltage drop is temperature sensitive (–2.2mv/°C). The circuit of Figure 6.1(b) uses an operational amplifier to eliminate this undesirable voltage drop and provide a high input resistance and low output resistance in the conducting state. In this circuit, diode D_1 blocks negative voltages from appearing across R_L.

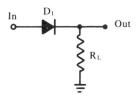

Figure 6-1a. Simple half wave rectifier.

Thus, for negative input voltages, the inverting terminal remains at ground and with the + input negative and diode D_1 blocking the feedback path, the operational amplifier output saturates at the negative supply. When the input goes positive, the high gain of the operational amplifier causes the diode to conduct when the + input is only a few microvolts more positive than the inverting terminal. For example, assuming that the diode conducts at 0.5 volts, with an open-loop gain of 100,000 (80dB), the operational amplifier. output voltage will be 0.5 volts when the input voltage differential is 0.5 volts/100,000 i.e. 5μ volts. Thus, because the inverting terminal is connected to the load resistor, R_L, the output "b" will be almost identical to the input "a" for positive inputs. A negative half-wave rectifier is created by simply reversing diode D_1.

The circuit of Figure 6-1(b) is the simplest precision rectifier connection, but it suffers from several limitations. First, during negative input excursions a very large input differential is seen by the input terminals as can be seen by comparing waveforms "a" and "b" of Fig. 6-1(b). This

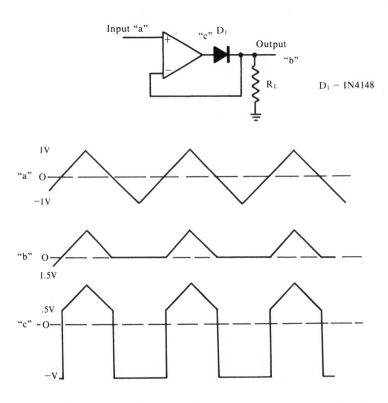

Figure 6-1b. Precision half wave rectifier and wave forms (voltage follower connection).

causes two problems: 1) an operational amplifier specified for large input voltage differentials must be used, thus eliminating many popular types such as the LM308, and 2) this large input differential causes all stages of the operational amplifier to saturate, and when the input goes positive, as much as 10-20μs may be required for the stages to return to their linear regions, placing severe frequency limitations on the rectifier's performance. Second, when the diode is conducting, the operational amplifier is connected in the voltage follower mode requiring that the operational amplifier be compensated for unity gain and thus further restricting the useful range of input frequencies. A circuit which overcomes these difficulties is shown in Figure 6-2.

In this circuit, an LM101 operational amplifier is used in the inverting connection. Again, D_1 blocks the operational amplifier from the load when "c" is negative ("a" positive) and when the input "a" is negative, the high gain operational amplifier effectively eliminates the diode forward drop (0.5 volts in this example). Diode D_2 clamps the output voltage to

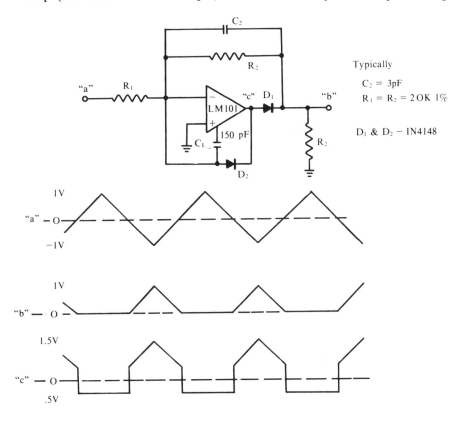

Figure 6-2. Precision half wave rectifier and wave forms (inverter connection).

$\simeq$–0.5 volts when "a" is positive providing two advantages, 1) the amplifier is maintained in the linear mode (non-saturated) at all times and 2) because saturation is avoided, the inverting input terminal is maintained at a virtual ground thus insuring that "b" will be at ground potential when "a" is positive. Because of this last point it can be seen that D_2 is not "optional," but is required for proper circuit operation. Because the amplifier is used in the inverting mode and linear stage operation is assured, feed-forward frequency compensation (see Chapter 1) can be used, thus greatly extending circuit bandwidth. Therefore, it is desirable to use an uncompensated operational amplifier with feed-forward capability (e.g. LM101, LM308) to maximize frequency response.

Even at low frequencies, say 1kHz or less, feed-forward compensation is desirable to maintain high precision since the ability of the operational amplifier to compensate for the diode drop is primarily a function of open-loop gain-bandwidth. With feed-forward compensation, circuit stability is achieved while increasing the open-loop gain-bandwidth characteristic at moderate frequencies as compared to standard unity gain frequency compensation connections.

Although the circuit of Figure 6-2 is shown with unity inverting gain, by making R_2 larger than R_1 and changing C_2, this circuit can provide rectification *and* gain.

Full Wave Rectifier

The circuit of Figure 6-3 is probably the simplest configuration which provides full wave rectification *and* a high resistance input with low resistance output. As shown in the wave forms of Figure 6-3 operational amplifier No. 1 operates as an inverting half-wave rectifier for positive voltages at "a" with diodes D_1 and D_2 functioning as previously described. Op. amp. No. 2 is used as a summing amplifier with inputs "a" and "b" weighted by resistor R_1 and R_2 respectively. With the input, "a", negative the voltage at point "b" is zero and the output "c" is just the input inverted as shown in period No. 1 Figure 6-3. During period No. 2 when the input is positive, the summing amp. has two non-zero inputs "a" and "b" which are equal in value, but opposite in polarity. Since $R_2=1/2R_1$, the *negative* voltage at "b" causes exactly twice as much current to flow *out* of the summing junction as the positive voltage at point "a" causes to flow *into* the summing junction. Feedback action of operational amplifier No. 2 causes "c" to go positive to balance the net currents *into* and *out of* the summing junction. With the values shown, resistors R_1 and R_2 carry equal currents into the summing node (virtual ground) and therefore the voltages at "a" and "c" are identical during period No. 2. All resistors must be carefully matched if the gains for positive and negative inputs are to be equal. Lacking

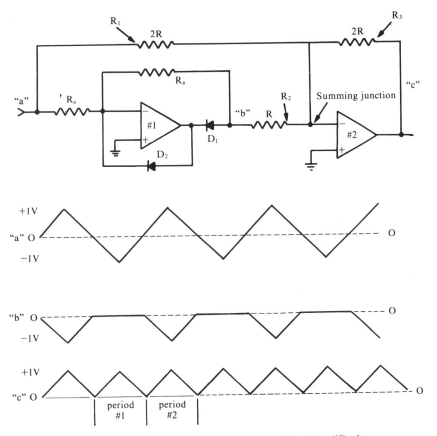

Figure 6-3. Precision full wave rectifier simplified.

matched resistors, R_2 could be constructed of a resistor and a potentiometer as shown in Figure 6-4. Several other improvements are included in the circuit of Figure 6-4. Of special note is the use of potentiometer R connected to point "a."

This resistor in conjunction with operational amplifier offset voltage nulling circuitry is required to achieve good performance with submillivolt inputs. Pot R functions as follows. For perfect nulling, zero input voltage should result in zero output voltage and this implies that point "a" must be zero also. However, with input bias currents flowing into both inverting inputs and diode D_1, preventing current from flowing from point "b" to point "a," it is impossible to set point "a" to zero. The positive current supplied to point "a" by R allows this voltage to be zeroed. With the circuit nulled at zero, *one-half* of the bias current used by *each* inverting input must be supplied from point "a" and therefore the value of R is just

$$R = \frac{V_{D3}}{I_{bias}} \approx \frac{0.6}{120nA} = 5 \text{ Meg} \qquad (6\text{-}1)$$

for a pair of typical LM101's. The diode voltage drop V_{D3} is temperature dependent and helps cancel the temperature dependence of the input bias currents.

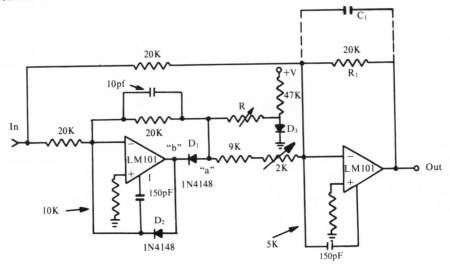

C_1 — used to filter output for A.C. to D.C. conversion

Figure 6-4. Fast full wave rectifier with bias current compensation, feed forward frequency compensation and gain control.

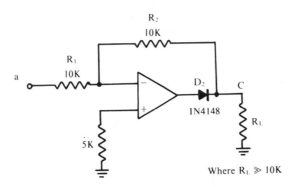

Where $R_L \gg 10K$

Figure 6-5b. Full wave rectifier using only one op amp.

If the full wave rectifier feeds a high resistance circuit such as a noninverting amplifier, the circuit of Figure 6-5 is useful if frequencies are low. For negative inputs, circuit operation is as previously described, but for positive inputs, D_1 blocks the operational amplifier output and, providing the operational amplifier can withstand large input differentials, the input "a" appears directly across R_L through the series combination of R_1 and R_2.

If R_L is fixed, R_2 can be adjusted to yield equal gains in the positive and negative directions. Under these conditions, R_2 will be slightly lower than R_1 depending on R_L. Clearly this circuit suffers in several areas: 1) it requires an operational amplifier which will stand large input differentials while maintaining low input currents; 2) all operational amplifier stages saturate during operation; 3) because the inverting terminal is not a virtual ground, feed-forward compensation can not be used and additional frequency limitations accrue; 4) the output resistance is low in one state and high in the other state. Otherwise, this circuit can be quite useful.

Full Wave Rectification Using Analog Gates

Most analog systems that use precision rectifiers require them so that some operations, say peak detection, analog division or analog to digital conversion, can be performed on a *single* signal polarity. In such systems, input signal polarity is measured with a comparator in front of the precision rectifier and carried through the system as a logic 1 (+) or 0 (−). In the case of an analog to digital converter, for example, the rectified signal would be sampled and applied to the input of the converter and following the conversion the digital polarity signal could be added as the most significant bit to the digital output. Such a system has two advantages: 1) a zero-based single polarity converter can be used on bipolar input signals; and 2) the addition of the sign bit to the MSB position effectively adds one bit of accuracy to the conversion, for example, an 8-bit converter would output a 9-bit number. Full bipolar analog division of two numbers represents another system that requires restriction of the input polarities.

The circuit of Figure 6-6 performs the polarity generation and rectification necessary for these systems and easily provides both positive and negative full wave outputs. Operation is simple, the LM308 unity gain inverter generates the inverse of the input signal. The CMOS analog switches, under control of the polarity indicator, select the real or inverted signal as required to perform negative or positive full wave operation. For example, if the input is positive, the control at A and B will be low selecting A_o and B_o inputs to appear at the com A and com B outputs respectively. Thus com A will be the inverted input and com B will represent the true input. For negative signals the selections are reversed and full wave rectification achieved. For high precision applications the inverter gain should be exactly 1 and its offset nulled. Since the 4051 are matched low resistance switches, excellent precision is maintained at the output. For detailed comparator application information, see Section 6.3. With a feed-forward compensated inverter this circuit achieves excellent high frequency operation because high open-loop gains are not required to mask diode voltage drops as in previously discussed designs.

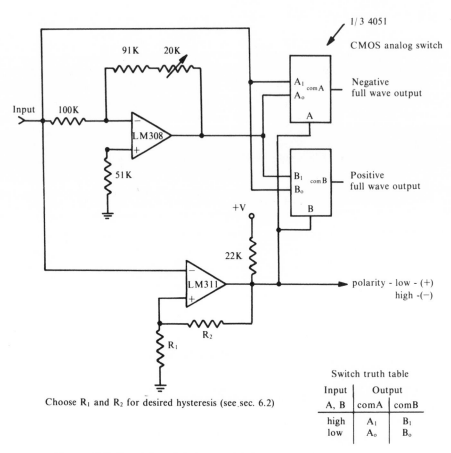

Figure 6-6. Precision full wave with input polarity indicator.

Switch truth table

Input	Output	
A, B	comA	comB
high	A_1	B_1
low	A_o	B_o

Choose R_1 and R_2 for desired hysteresis (see sec. 6.2)

6.2.2 A.C. to D.C. Converters

Any of the rectifiers presented in Section 6.2.1 could be used as an A.C. to D.C. converter just by low-pass-filtering the output voltages. With the circuit of Figure 6-4 this filtering is especially easy to perform by the addition of a single capacitor C_1. The 3 dB down point of this filter is

$$f_{3dB} = \frac{1}{2\pi R_1 C_1} \tag{6-2}$$

For effective filtering, the minimum input frequency should be at least $10 \times f_{3dB}$.

6.2.3   Peak Detectors

As was the case with the simple rectifier of Figure 6-1(a), the simple peak detector, Figure 6-7(a) suffers an inaccuracy due to the diode voltage drop. Again, an operational amplifier can be used to virtually eliminate this effect. In the circuit of Figure 6-7(b) as long as the input voltage is increasing, diode D_1 is conducting, the amplifier is in the voltage follower mode and e_{out} follows the input. Since D_1 blocks current from flowing out of the capacitor, e_{out} is always the maximum value that the input achieved. Resistor R is included to remove the capacitive load from the operational amplifier at high frequencies when the diode is conducting. If R were not present, an additional high frequency pole would be created in the feedback loop by the operational amplifier output resistance and capacitor C, resulting in ringing or oscillations even with unity gain compensation.

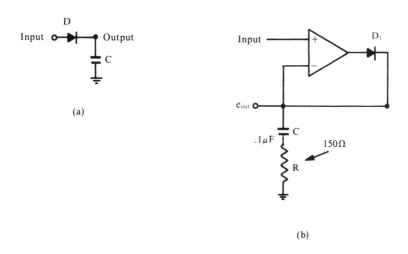

Figure 6-7. Improved peak detector using an op amp.

If a low leakage capacitor is used for C (see Chapter 2 Section 2.3), the output voltage, e_{out}, will decay at a rate determined by the input bias current I_b requirements of the operational amplifier and the leakage current I_L of D_1. Assuming that both of these currents flow out of the capacitor, the drift rate is given by

$$\frac{\Delta e_{out}}{\Delta t} = \frac{I_L + I_b}{C} \text{ V/sec} \qquad (6\text{-}3)$$

For example, if $I_b = 20nA$ and $I_L = 5nA$ and $C = 0.1\mu F$, typical values for a circuit using an LM101, then

$$\frac{\Delta e_{out}}{\Delta t} = \frac{25 nA}{0.1 \mu F} = 0.25 \ V/sec$$

For some applications this would be an unacceptably high discharge rate. Equation 6-3 states that the drift rate can be lowered by raising C or lowering the leakage currents. As a practical matter C should not be increased above $1 \mu F$ since low leakage capacitors above this value are large and expensive. Leakage currents, I_L and I_b, can be reduced by specifying FET input op. amps. and high quality diodes.

The major shortcoming of the circuit of Figure 6-7(b) is that e_{out} is not buffered. To utilize the information stored on C requires an additional sensing circuit which adds another leakage path. A circuit which provides low drift and output buffering using inexpensive bipolar op. amps. is shown in Figure 6-8(a). This design uses two low leakage, epoxy-cased FET's Q1 and

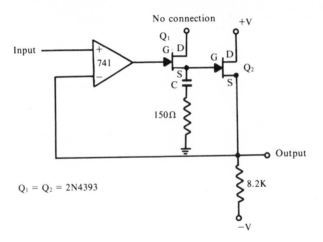

Figure 6-8a.

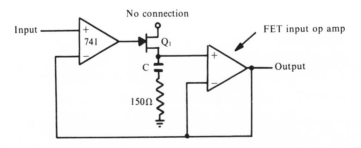

Figure 6-8b. Buffered low drift peak detectors.

Q_2 and a 741 operational amplifier. Q_1 is used as a low leakage diode (a technique applicable to previously discussed circuits) and Q_2 is used as a buffer in the source-follower mode. Because the output is connected back to the inverting input the high open-loop gain of the op. amp. will correct for both the diode drop and the gate-to-source voltage of Q_2, ensuring that the output will track the input in the positive direction. One subtle advantage of this connection is that when the input is less than the output and diode Q_1 is reversed biased, the diode leakage current flows *out* of the capacitor, but the gate leakage of the source follower Q_2 flows *into* the capacitor and these effects tend to cancel. Of course, the source follower, Q_2, could be replaced by a high quality FET input operational amplifier connected in the voltage-follower mode as shown in Figure 6-8(b). Output voltage drifts can be reduced below 1 mV./sec with either connection.

6.2.4   Sample and Hold Circuits

Sample and hold circuits are usually required at the input of an analog-to-digital converter. Some types of converters, e.g. multislope, do not need a sample-hold for accurate operation but for other types, e.g. successive approximation, a sample-hold is mandatory (see Chapter 10).

Figure 6-9 shows a very simple sample-hold circuit using either an MOS enhancement mode FET or a junction FET. In this design, the operational amplifier must be compensated for unity gain.

As shown, the LM311 driver can be controlled by either CMOS or TTL logic signals. (For an explanation of the LM311 output stage see Section 6.3). The "ground" terminal of the LM311 is connected to the negative supply voltage so that the control signal for either switch goes to the negative supply in the off (hold) mode. In the sample mode the control signal for the enhancement-mode device, Q_1, swings to the positive supply giving this circuit a dynamic input range which includes the negative supply and reaches within a few volts of the positive supply. The junction FET, Q_2, control signal in the sample mode simply allows the gate and source to be at the same potential, ensuring high channel conductivity. This circuit is limited in the positive direction only by the common-mode capability of the operational amplifier.

In the hold mode, voltage drift is a function of switch leakage I_s and the input bias current requirements of the amplifier, I_b. The voltage drift in volts/sec is given by

$$\frac{\Delta V}{\Delta t} \; = \; \frac{I}{C} \; = \; \frac{I_s + I_b}{C} \;\; V/sec \qquad\qquad \textbf{(6-4)}$$

assuming that the storage capacitor C is a low leakage film type.

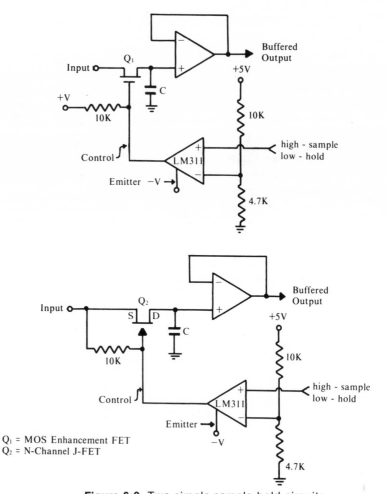

Figure 6-9. Two simple sample-hold circuits.

The ability of these circuits to respond to rapid input changes in the sample mode is limited by the filter consisting of the switch (Q_1 or Q_2) on resistance R_s; and the storage capacitor C so that the signal will be 3dB down at the operational amplifier input as determined by

$$f_{3dB} \quad \frac{1}{2\pi R_s C} \qquad \qquad (6\text{-}5)$$

Of course, for high accuracy, the input frequency must be significantly lower than determined by Equation 6-5. By comparing Equation 6-4 and Equation 6-5, we see that fast response and low drift are conflicting requirements. Best performance is obtained with a very low input bias current operational amplifier and a switch with a low ON resistance *and* low leakage.

Higher accuracy can be obtained by including the capacitor in a feedback loop as shown in Figure 6-10. The inputs and outputs of these circuits are fully buffered. Again, resistor R may be required for stability.

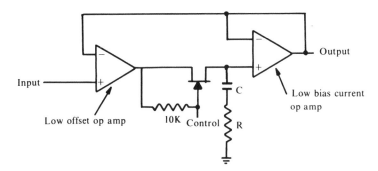

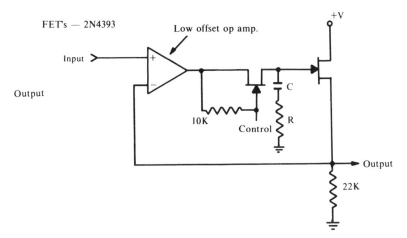

Figure 6-10. Fully buffered sample holds.

Achieving Capacitor Limited-Hold Performance

In the preceding designs for sample-hold circuits and peak detectors the basic limitations on storage time were the leakage currents of the diode or switch and the input bias requirements of the buffering operational amplifier. Presumably, the best a sample-hold circuit designer can do is to have the hold-mode decay rate determined only by capacitor leakage currents. That such performance is difficult to achieve can be appreciated by calculating the value of the leakage current for a high quality film capacitor. For example, consider an ARCO Electronics Inc., Type PC polystyrene capacitor specified with a megohm-microfarad product (see Chapter 2 Section 2.3) of 10^6. For a1 μF capacitor the insulation resistance would be

$$\frac{10^6 \text{ megohm} \times \mu F}{1 \mu F} = 10^6 \text{ megohm}$$

With a 5-volt charge on this capacitor the leakage current is just

$$\frac{5 \text{ volts}}{10^6 \text{ megohm}} = 5 \text{ pA}.$$

It is possible to purchase an operational amplifier with sub pico amp input bias current, for example, the National LH0052 FET input operational amplifier. However, no switch exists with sub pico amp leakage current specifications. The circuit of Figure 6-11 circumvents the switch leakage problem using only one expensive component — the LH0052. The CMOS analog switch 4051 connects the common output, com X, to one of two inputs, X_1 or X_o, depending on the state of the control, X, and is used to siwtch the circuit into two states—"Sample" or "Hold." In the sample mode Figure 6-11b, the high open-loop gain of the 741 input operational amplifier eliminates the voltage drop of diodes D_1 and D_2 and because the overall loop is closed due to the connection to the inverting terminal of 741, the capacitor is charged to the input voltage. Now, in the hold mode the LH0052 output is connected back to the diodes as shown in Figure 6-11c. With the low offset voltage of the LH0052, less than a millvolt will be across diodes D_1 and D_2 since the capacitor or voltage is on one side of the diodes and the LH0052 output is on the other side. Under these conditions virtually no current flows in either diode and therefore, capacitor-limited performance is achieved with inexpensive switches. As mentioned, a resistor may be required in series with C to achieve circuit frequency stability.

The rate of discharge of this circuit is given by

$$\frac{\Delta V}{\Delta t} = \frac{I}{C} = \frac{5pA}{1\mu F} = 5\mu V/\text{sec}$$

However, since the capacitor leakage is responsible for this drift, the actual rate will be a function of the storage voltage. The above calculation assumes a 5-volt charge on the capacitor. Clearly, this technique is applicable to peak detectors or any other capacitor storage circuit.

No-drift Sample-Hold

It should be remembered that the ultimate *no*-drift sample-and-hold circuit is an analog-to-digital converter. With the voltage stored in binary form in a digital register, drift is not a consideration. The voltage can be read out by a digital-to-analog converter. Such circuits find utility in machine control systems. (See Chapter 10 for typical A-D-A circuits.)

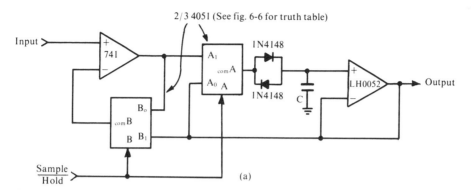

2/3 4051 (See fig. 6-6 for truth table)

(a)

Precision high speed low drift sample hold

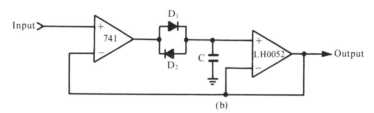

(b)

Equivalent circuit in sample mode

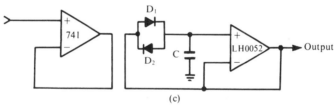

(c)

Equivalent circuit in hold mode

Figure 6-11. Achieving capacitor limited hold performance.

6.3   COMPARATOR APPLICATIONS

Voltage comparators are a special class of integrated circuits with input characteristics similar to op. amps., but with switching type outputs. As a circuit element, their function is to determine which of two input voltages is greater and display the results of this decision by setting the output high or low as shown in Figure 6-12. This same operation could be performed by an operational amplifier and in fact a comparator is a special type of operational amplifier circuit optimized for fast recovery from saturation. In addition, voltage comparators usually have special output

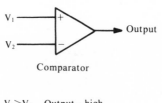

Comparator

$V_1 > V_2$ Output - high
$V_1 < V_2$ Output - low

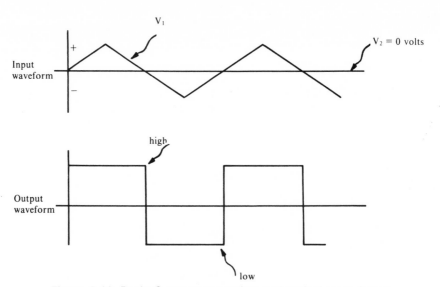

Figure 6-12. Basic Comparator action and typical wave forms.

stages to increase their versatility and compatability with digital logic gates. In many cases where speed is not a requirement, op. amps. can be used as voltage comparators. Voltage comparator inputs have specifications very similar to op. amps. Low offset voltage, low input bias current, wide common-mode range and the ability to withstand large voltage differentials are definite input requirements. An additional specification is the delay time, which measures the time required for the output to change states following a change in the input state. For a general purpose comparator like the LM311, the delay time is about 200 nanoseconds, whereas a fast ECL type comparator, e.g. the AM685, has a delay time of about 3 nanoseconds.

Several basic comparator applications are covered below. In these examples, the comparator output is assumed to switch between the plus and minus supply voltages.

Level Detector with Hysteresis

Figure 6-13a shows the basic level detection circuit using a comparator or an operational amplifier, depending on speed requirements. The resistor network R_1 and R_2 connected to the noninverting terminal provides positive feedback resulting in the hysteresis transfer function shown. To understand the purpose of this network, let us consider a typical output transition. Assume that the output is high (+15 volts). Therefore, the voltage at the noninverting terminal, V+ will be higher than the reference because of the resistor network. Specifically, when the output is high,

$$V_+^H = V_{ref} + (15V - V_{ref}) \frac{R_1}{R_2 + R_1} \qquad (6\text{-}6)$$

If the input slightly exceeds V^H, the output will switch low (–15 volts) forcing V+ to a slightly lower voltage given by

$$V_+^L = V_{ref} - (V_{ref} - 15) \frac{R_1}{R_2 + R_1} \qquad (6\text{-}7)$$

Therefore, the actual threshold voltage depends on the output state. If the input crosses the threshold in the positive direction, curve No. 1, the threshold is lowered by the output transition and vice versa (see curve No. 2). For slowly varying input signals, this positive feedback has the effect of decreasing the delay time of the comparator by generating a larger difference signal at the input during the transition. This effect can be enhanced by the addition of C, Figure 6-13a, without upsetting the d.c. accuracy. A second advantage of adding hysteresis is increased tolerance to noise voltages on the input signal. Without hysteresis these noise voltages cause multiple switching to occur on slow input transitions. When noise voltages are a problem it is desirable to set the total hysteresis voltage larger

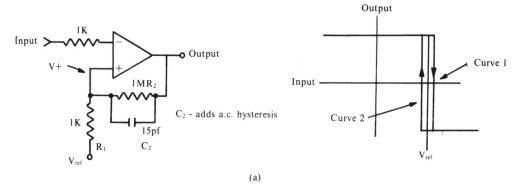

(a)

Figure 6-13a. Level detector with hysteresis.

than the magnitude of the expected noise. The total hysteresis voltage is given by

$$V_{hys} = (V^+ - V^-) \times \frac{R_1}{R_1 + R_2} \qquad \text{(6-8)}$$

where V+ and V– are the comparator output voltage levels in the high and low state respectively. Note that the reference voltage does not appear in Equation 6-8. Typically, the hysteresis voltage, V_{hys}, is set at about 5-20 mv.

Zero Crossing Detector

Figure 6-13(b) shows a zero crossing detector which gives an output when the input is within a band about zero with upper bound V+ determined by R_1 and R_2, and lower bound V– determined by R_3 and R_4. If the input is outside this band, then the output of either comparator No. 1 or No. 2 is low, which holds the output low through one of the switching diodes. When the input is inside the band, both comparators are high and the circuit output is pulled high by R_5. Thus, the resistor diode network at the output performs a logic "and" function.

Using various networks V+ and V– could be set at any value as long

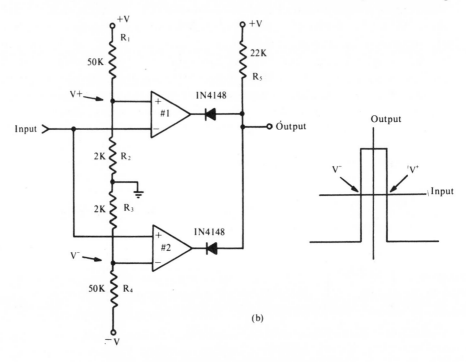

(b)

Figure 6-13b. Zero crossing detector.

as V+ > V−. In this mode the circuit becomes a double-ended limit detector which yields an output when the input is between V+ and V−. This circuit is useful in automated testing procedure where it is desirable to know if a measured parameter, e.g. the value of a resistor, falls within a given range.

Other Applications

The majority of comparator applications occur in conjunction with other circuit elements. See Chapter 4 for oscillator circuits which use comparators and see Chapter 10 which explores several analog-to-digital converter circuits utilizing voltage comparators.

Comparator Output Stages

a) *Dedicated output stages.* One class of comparators has an output stage intended for operation with a specific logic family. For example, the 710 and AM 586 have TTL output stages and the AM 585 has an ECL output stage.

b) *Open collector.* Several comparators offer an open collector output. As shown in Figure 6-14(a) the output is connected through a resistor to a voltage which is greater than the negative supply voltage of the comparator. Therefore, the high level is controlled by the value of V, but the low level is restricted to the negative supply voltage. The LM 339 is an example of such a comparator.

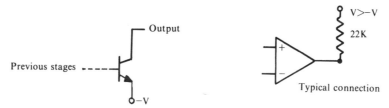

Figure 6-14a. Open collector output.

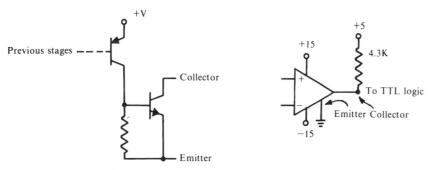

Figure 6-14b. Floating transistor output.

c) *Floating output transistor.* With the floating transistor structure of Figure 6-14(b) maximum versatility is obtained since both the low and high output voltages can be set independent of supply voltage. In the application shown, the comparator is operated from + and − 15 volt supplies, thereby achieving wide input common mode range while interfacing directly to TTL logic gates. The LM 119 and LM 111 comparator families utilize this output structure.

Wire "or" Connection

Both the open collector structure and the floating transistor structure can take advantage of the wired "or" connection shown in Figure 6-14(c). Although a positive logic "and" gate is shown in the equivalent circuit it is an "or" gate for negative logic. Since the output structure can only sink current both comparators must be off for the output to go high. Conversely, either of the comparators could pull the output low. Clearly, a large number of comparators could be connected in this fashion.

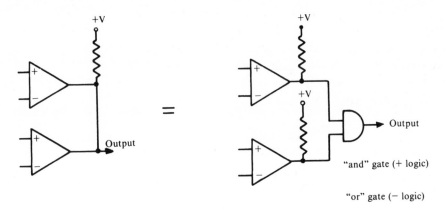

Figure 6-14c. Wired "or" connection.

Chapter 7

Active Filters

7.1 Filters are used to allow signals in certain frequency ranges to be transmitted through a system while preventing the transmission of signals in other frequency ranges. The advent of I.C. operational amplifiers, and in particular, the packaging of several operational amplifiers in one small package, make the construction of active filters simple and practical. The advantage of active filters in comparison with passive filters is that bulky RLC networks are replaced with small operational amplifiers and RC networks. This makes the design and construction of high quality filters attractive. In this chapter we show how to design simple and complex filters to meet specified signal transfer characteristics.

7.2 PRELIMINARY CONSIDERATIONS

The ideal transfer characteristics of *low-pass, high-pass*, and *band-pass* filters are shown in Figure 7-1. Ideally, the signal is not attenuated at all in the pass band of a filter, and is attenuated completely in the stop band. Also, ideally the phase shift of the signal is linear with frequency in the pass band. The phase shift in the stop band has no significance since the signal is not transmitted in this band. Under these ideal conditions the signal is transmitted without any distortion in the pass-band frequency range, and it is not transmitted at all outside this frequency range. In practice, the ideal filter can only be approached to various degrees, depending upon the complexity of the filter.

Filter Order

The simplest filter contains *one* capacitor for each cutoff frequency. This type is called a *first-order* filter. Thus a first-order low-pass filter

229

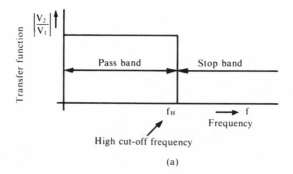

(a)

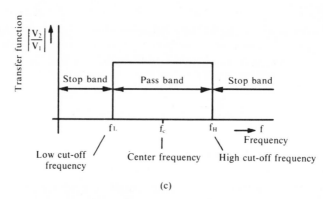

(b)

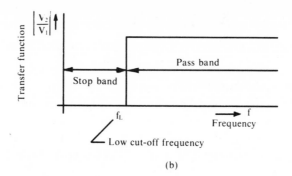

(c)

Figure 7-1. Transfer functions of ideal filters.
(a) Low pass filter
(b) High pass filter
(c) Band pass filter

contains one capacitor; a first-order high-pass filter contains one capacitor; and a first-order band-pass filter contains two capacitors. The next degree of complexity is obtained in the *second-order* filter, then the *third-order*, *fourth order*, etc. The higher the order of the filter, the more nearly does the filtering characteristic of the filter approach that of the ideal filter.

Filter Classes

A filter of any *given order* can be made to approximate the ideal filter in different ways depending on the values given to the filter components. The type of approximation chosen determines the filter *class.* Two useful classes are the *maximally flat* (also known as Butterworth) and the *equal-ripple* (also known as Chebishev) classes. The difference between the two classes is illustrated in Figure 7-2. Compare, for example, the gain-frequency characteristic of the second-order maximally flat band-pass filter given in Figure 7-2c (n=2) with this of the second-order, equal-ripple, band-pass filter given in Figure 7-2e. The maximally flat filter approximates the ideal filter better in the pass band, but attenuates signals in the stop band less than the equal-ripple filter. In Figure 7-2 only band-pass diagrams for equal-ripple filters are shown. Low-pass and high-pass filter characteristics can be deduced by considering the appropriate portion of the corresponding band-pass filter characteristic.

The particular order and class chosen for any given design depends on the particular system specifications. The design is usually carried out in terms of steady state frequency response specifications. However, the same filter components which determine the frequency response also determine the transient response. The mathematical relationships between the frequency response and the transient response are quite complex and some experimentation in practical design is necessary. The more "squared-off" the frequency response is, the more overshoot is encountered during a transient period. The frequency response curve of a fourth-order, low-pass maximally flat filter is compared with the response curve of a corresponding equal ripple filter in Figure 7-3.

Normalized filters

Filter design is simplified, particularly when designing higher-order filters, if the "normalized filter" concept is used. One first uses a simple set of design equations or tables to obtain the normalized filter. This filter has a cutoff frequency of 1 rad/sec (or $1/2\pi$=0.159 Hz) and usually resistors of 1 Ω and/or capacitors of 1 F. Two additional design steps result in the desired practical filter: first a change of component values (resistors or capacitors) is made to obtain the *desired* cutoff frequency, and finally a change in values of resistors and capacitors is made in a prescribed manner to obtain realistic orders of magnitude of component values. The values resulting from the computations are given to accuracies considerably beyond the tolerances of practical components. In practice, one uses the closest commercially available components and checks to see whether the filter characteristics are still acceptable and, if necessary, adds either resistor trimmers or capacitor trimmers to *tune* the filter to the exact desired cutoff frequencies.

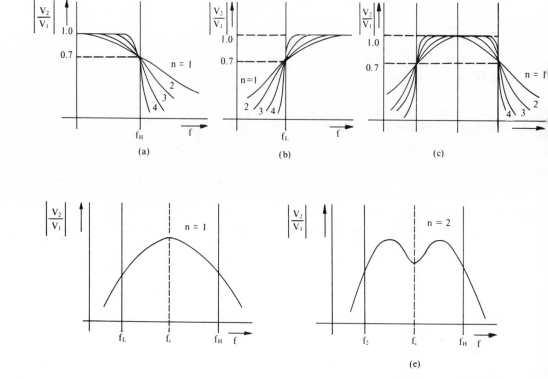

(a)

(b)

(c)

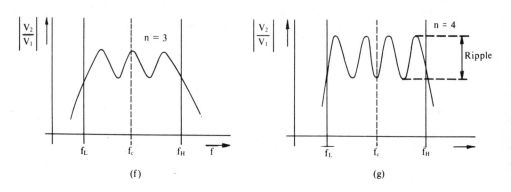

(e)

(f)

(g)

Figure 7-2. Gain transfer characteristics for *maximally flat* (Butterworth) and *equal ripple* (Chebishev) filters. (Not to scale.)

(a) Low pass maximimally flat
(b) High pass maximally flat
(c) Band pass maximally flat
(d) Band pass euqal ripple, n = 1
(e) Band pass equal ripple, n = 2
(f) Band pass equal ripple, n = 3
(g) Band pass equal ripple, n = 4

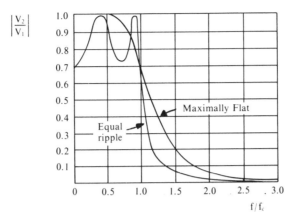

Figure 7-3. Response curves of magnitudes of low-pass fourth order *maximally flat* (Butterworth) and *equal ripple* (Chebishev) filters.

7.3 FILTER DESIGN

We shall now give practical filter configurations and design procedures. Loose specifications can be met with simple first-order filters. Since the circuits are simple and the number of components is small it is possible to design a first-order filter directly without the need of the intermediate normalized filter step. Somewhat tighter specifications can be met with second-order filters. Third-order filters find wide use in a large number of practical applications. Tighter specifications require fourth-order filters. The details of the specifications, which may include the percentage of permissible ripple in the pass band, the cutoff frequency and the required attenuation at specific frequencies in the stop band, make the generalized design approach which uses the normalized filter concept necessary to effect a simple easy-to-follow design procedure for higher-order filters.

7.3.1 First-Order Filters

Low-Pass Filter

The circuit diagram of a first-order low-pass filter is shown in Figure 7-4. The 3 dB cut off frequency is given by

$$f_H = \frac{1}{2\pi R_2 C_2}$$

The d-c gain is R_2/R_1, and the roll-off in the stop band is –20 dB/dec (–6 dB/oct).

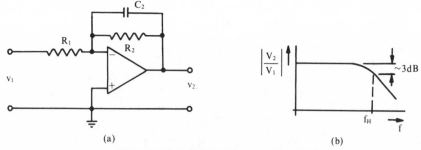

Figure 7-4. First-order low-pass filter
(a) Circuit diagram
(b) Frequency characteristic

Example: For a low-pass amplifier with f_H = 1000 Hz and d-c gain = 10,

$$\frac{1}{2\pi R_2 C_2} = 1000$$

$$\frac{R_2}{R_1} = 10$$

We have two equations with three unknowns. We can therefore choose arbitrarily any one of the three components. Let C_2 = 0.01 μF. From the equations above we obtain R_2 = 15.9 KΩ, and R_1 = 1.59 KΩ.

High-Pass Filter

The circuit diagram of a first-order high-pass filter is given in Figure 7-5. The 3 dB cutoff frequency is given by

$$f_L = \frac{1}{2\pi R_1 C_1}$$

and the high frequency gain beyond the cutoff frequency is R_2/R_1. The roll-off below the cutoff frequency is 20 dB/dec (6 dB/oct).

Example. For a high-pass amplifier with f_L = 20 KHz and high frequency gain of 10,

$$\frac{1}{2\pi R_1 C_1} = 20,000$$

$$\frac{R_2}{R_1} = 10$$

Let C_1 = 10,000 pF. From the equations above R_1 = 796Ω and R_2 = 7.96 KΩ.

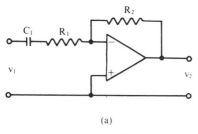
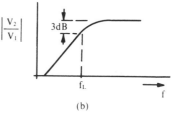

Figure 7-5. First-order high-pass filter
(a) Circuit diagram
(b) Frequency characteristic

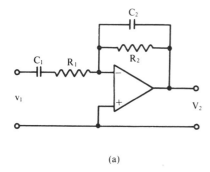
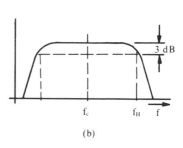

Figure 7-6. Simple band-pass filter.
(a) Circuit diagram
(b) Frequency characteristic

Band-Pass Filter

The circuit diagram of a first-order band-pass filter is shown in Figure 7-6. The low and high cutoff frequencies are, respectively,

$$f_L = \frac{1}{2\pi R_2 C_2} \qquad \text{and } f_H = \frac{1}{2\pi R_1 C_1}$$

The mid-frequency gain is R_2/R_1. There are two capacitors in this circuit, one associated with the low cutoff frequency and one with the high cutoff frequency. The slopes of the transfer curve below f_L and above f_H are 20 dB/dec (6 dB/oct) and –20 dB/dec (6 dB/oct), respectively.

Example. It is desired to design a filter with f_L = 2 kHz, f_H = 5 kHz and mid-frequency gain = 5. The design equations are

$$\frac{1}{2\pi R_2 C_2} = 2,000$$

$$\frac{1}{2\pi R_1 C_1} = 5{,}000$$

$$\frac{R_2}{R_1} = 5$$

We have three equations with four unknown parameters. We can, therefore, choose any one of the four arbitrarily. Let $C_2 = 0.01\ \mu F$, then $R_2 = 7.958\ K\Omega$, $R_1 = 1.592\ K\Omega$, and $C_1 = 0.02\ \mu F$.

7.3.2 Second-Order Filters

Low-Pass Filter

A second-order low-pass filter is shown in Figure 7-7. The gain of the amplifier is $K = 1 + R_b/R_a$.

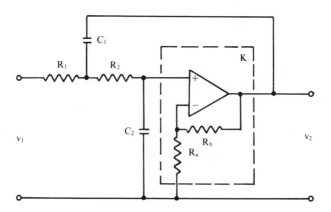

Figure 7-7. Second-order low-pass filter.

The transfer function of the filter is determined by the value of the amplifier gain and by the values of the resistors and capacitors. The same transfer function can be realized with different combinations of these values. Three choices are particularly useful: (a) $K = 2$. This results in $R_a = R_b$ which gives good matching and tracking; (b) $K = 1$. This results in a voltage-follower amplifier, and eliminates the need for the two resistors R_a and R_b: R_a is replaced with an open circuit and R_b with a short circuit; (c) $R_1 = R_2$ and $C_1 = C_2$. This again results in good matching and tracking. Further, whenever equal components are used, there will be a lower cost when a large number of filters is built. The filter design is first carried out for the cutoff frequency $\omega_H = 1$ rad/s. The component values are then changed in two steps: the first step gives the correct cutoff frequency, and the second step gives reasonable values of resistors and capacitors. Table 7-1 lists gain and

Filter Class	R_1	R_2	C_1	C_2	K
Maximally flat	1.00000	1.00000	0.87403	1.14412	2.00000
(Butterworth)	1.00000	1.00000	1.41421	0.70711	1.00000
3.01 dB at ω_H	1.00000	1.00000	1.00000	1.00000	1.58578
Equal ripple	1.00000	1.00000	0.77088	0.855557	2.00000
(Chebishev)	1.00000	1.00000	1.40259	0.47013	1.00000
0.5 dB ripple	0.81220	0.81220	1.00000	1.00000	1.54213
Equal ripple	1.00000	1.00000	0.93809	0.96688	2.00000
(Chebishev)	1.00000	1.00000	1.82192	0.49783	1.00000
1 dB ripple	0.95237	0.95237	1.00000	1.00000	1.95446

Table 7-1. Component and gain values for the low-pass filter of
Figure 7-7 for $\omega_H = 1$)
(R in ohms; C in farads)

component values for the three choices for $\omega_H = 1$ for the maximally flat filter
and for two equal ripple filters, one with 0.5 dB ripple, and the other with 1
dB ripple.

Example. Design a low-pass, zero ripple, filter with a cutoff
frequency of 1500 Hz and a d-c gain of 2.

Solution. The normalized filter is a maximally flat filter with a circuit
configuration of Figure 7-7 and values $R_1 = R_2 = 1.00000$ ohms, $C_1 = 0.87403$
F, and $C_2 = 1.1442$ F, as obtained from Table 7-1. To obtain a cutoff
frequency of $\omega_H = 2 \pi 1500$, we divide all capacitors by $2 \pi 1500$ giving

$$C_1 = 0.87403/2 \pi 1500 = 92.7375 \ \mu F$$

$$C_2 = 1.14412/2 \pi 1500 = 121.3949 \ \mu F$$

Finally, to obtain a set of reasonable values, we multiply the resistors
by a factor of 10^5 and divide the capacitors by a factor of 10^5 (provided the
factors are the same, the filter characteristics will not change). For good
economy we make $R_a = R_b = R_1 = R_2$. The component values of the filter are:

$$R_1 = R_2 = R_a = R_b = 100 \ \text{Kohms}$$

$$C_1 = 927.375 \ \text{pF}$$

$$C_2 = 1213.949 \ \text{pF}$$

Note: The designer might prefer to choose resistors of values of 10 K ohms to reduce bias current-offset effects, depending on the particular amplifier used (see Chapter 1). In this case the capacitors must be multiplied by a factor of 10, giving $C_1 = 0.00927 \ \mu F$ and $C_2 = 0.01214 \ \mu F$.

In practice, the designer must use commercially available capacitors. A choice of a ceramic capacitor of $0.01 \ \mu F$ for C_1 and $0.01 \ \mu F$ or $0.015 \ \mu F$ for C_2 is suitable. Since these are not exactly equal to the calculated values, the cutoff frequency will be slightly different from the one specified. Rarely is this deviation of any consequence. If for some reason the precise cutoff frequency must be obtained, it is necessary to shunt the capacitors with trimming capacitors. In this case the fixed capacitors must have smaller values than the ones computed, e.g. $0.008 \ \mu F$ for C_1 and $0.01 \ \mu F$ for C_2 and the trimmer capacitors will be adjusted to add the required amount of capacitance. If a precise cutoff frequency is required, trimmers (capacitor trimmers or resistance trimmers) must be added anyway since commercial capacitors have typically at best $\pm 20\%$ tolerance.

High-Pass Filter

The circuit diagram of a high-pass filter is given in Figure 7-8. Component values for the normalized filter are listed in Table 7-2. To see how to make use of this information, the reader must refer to the section on low-pass filters, and follow the same procedure. Select components from the Table for the desired filter class, for $\omega_L = 1$, divide all capacitor values by the desired cutoff frequency; multiply all resistance values and divide all capacitance values by a suitable factor to obtain reasonable practical values of components and round the values off to commercially available values; use trimmers if necessary.

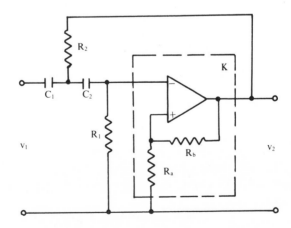

Figure 7-8. Second-order high-pass filter.

Filter Class	R_1	R_2	C_1	C_2	K
Maximally flat	1.00000	1.00000	1.41421	0.70711	2.00000
(Butterworth)	0.70711	1.41421	1.00000	1.00000	1.00000
3.01 dB at ω_1	1.00000	1.00000	1.00000	1.00000	1.58579
Equal ripple	1.00000	1.00000	1.42563	1.06356	2.00000
(Chebishev)	0.71281	2.12707	1.00000	1.00000	1.00000
0.5 dB ripple	1.23134	1.23134	1.00000	1.00000	1.8422
Equal ripple	1.00000	1.00000	1.09772	1.00436	2.00000
(Chebishev)	0.54586	2.00872	1.00000	1.00000	1.00000
1 dB ripple	1.05001	1.05001	1.00000	1.00000	1.00000

Table 7-2. Components and gain values for the high-pass filter of Figure 7-8 for $\omega_1 = 1$
(*R* in ohms; *C* in farads)

Band-Pass Filter

The circuit diagram of a band-pass filter is shown in Figure 7-9. A band-pass filter is characterized by the center frequency ω_C and by the Q of the filter. The Q of a filter is defined as the ratio of f_c (the center frequency) to the bandwidth (the difference $f_H - f_L$). The gain at the center frequency is given by

$$\frac{KQ}{R_1 C_1}$$

Table 7-3 lists values of K and of components for four different values of Q for the normalized (i.e., $\omega_c = 1$) filter. The following example demonstrates how to design a band-pass filter of the configuration of Figure 7-9 with the aid of Table 7-3.

Example. Design a band-pass filter with a center frequency of 10 KHz and a Q of 5.

Solution. The circuit diagram is shown in Figure 7-9. From Table 7-3 we select for Q = 5 the values:

$$R_1 = 1,0000, R_2 = 0.63439, R_3 = 2.57630$$
$$C_1 = C_2 = 1.00000, K = 2.00000$$

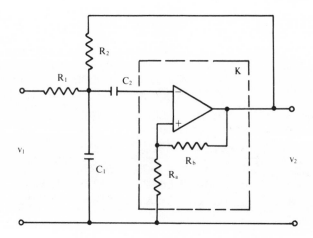

Figure 7-9. Second-order band-pass filter.

Q	R_1	R_2	R_3	C_1	C_2	K
2	1.41421	1.41421	1.41421	1.00000	1.00000	3.29284
	1.00000	0.74031	2.35078	1.00000	1.00000	2.00000
5	1.41421	1.41421	1.41421	1.00000	1.00000	3.71716
	1.00000	0.63439	2.57630	1.00000	1.00000	2.00000
10	1.41421	1.41421	1.41421	1.00000	1.00000	3.85858
	1.00000	0.60471	2.63587	1.00000	1.00000	2.00000
20	1.41421	1.41421	1.41421	1.00000	1.00000	3.92428
	1.00000	0.59076	2.69274	1.00000	1.00000	2.00000

Table 7-3. Component values for the band-pass filter of Figure 7-9 for $\omega_c = 1$ rad/sec.
(R in ohms; C in farads)

The value of K gives the gain of the amplifier:

$$K = 1 + \frac{R_a}{R_b}$$

We choose $R_a = R_b = 1.00000$ giving K = 1+1 = 2

The circuit of Figure 7-9 with component values listed above gives a filter with the desired Q = 5, but a center frequency ω_c = 1. To obtain a center frequency f_c = 10 kHz (or ω_c = 2 π 10,000), we divide the capacitor values by 2 π 10,000.

$$C_1 \quad = \quad C_2 \quad = \quad \frac{1}{2\ \pi\ 10,000} \quad = \quad 15.9\ \mu F$$

Finally, to obtain reasonable values, we multiply all resistors by 10^4 and divide all capacitors by 10^4. The result is:

$R_1 = R_a = R_b = 10\ K\Omega;\ R_2 = 6.34\ K\Omega;\ R_3 = 25.8\ K\Omega$
$C_1 = C_2 = 1590\ pF$

The nearest standard commercial values are:

$R_1 = R_a = R_b = 10\ K\Omega;\ R_2 = 6.2\ K\Omega;\ R_3 = 27\ K\Omega$
$C_1 = C_2 = 1500\ pF$

These are slightly different from the computed values causing f_c and Q to be slightly different from the specified values. If exact specifications must be met, trimmer capacitors must be added in shunt (parallel) with C_1 and C_2.

7.3.3 Third-Order Filters

Good characteristics for low-pass and high-pass filters for a wide range of applications can be obtained with third-order filters. These are perhaps the most popular general purpose filters. We shall present design procedures for maximally flat and a range of equal-ripple filters.

Low-Pass Filter

The circuit diagram of a low-pass third order filter is shown in Figure 7-10. Component values to achieve particular characteristics are listed in Table 7-4. The maximally flat filter frequency response curve has an asymptotic slope in the stop band of 60 dB/dec (18 dB/oct). The exact attenuation at any frequency ω is given by

$$\frac{V_2}{V_1} = \frac{1}{(1 + \omega^6)^{1/2}} \tag{7-1}$$

The exact attenuation for the equal-ripple filters in the stop band is given by

$$\frac{V_2}{V_1} = \frac{1}{[1 + \epsilon^2(4\omega^3 - 3\omega)^2]^{\frac{1}{2}}} \tag{7-2}$$

where ϵ^2 is the *ripple factor*. Values of the ripple factor for the ripple values in Table 7-4 are given in Table 7-5.

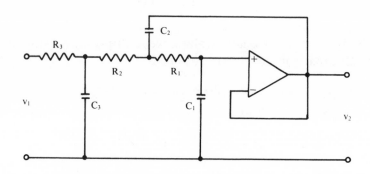

Figure 7-10. Third-order low-pass filter.
(Reprinted from *Electronics*, November 11, 1976; Copyright © McGraw-Hill, Inc., 1976)

Filter Class	R_1	R_2	R_3	C_1	C_2	C_3
Maximally flat (Butterworth) 3.01 dB at ω_2	1.00000	1.00000	1.00000	0.20245	3.5465	1.3926
Equal ripple (Chebishev) Ripple (dB)						
0.001	1.00000	1.00000	1.00000	0.07130	2.5031	0.8404
0.03	1.00000	1.00000	1.00000	0.07736	3.3128	1.0325
0.10	1.00000	1.00000	1.00000	0.09691	4.7921	1.3145
0.30	1.00000	1.00000	1.00000	0.08582	7.4077	1.6827
1.00	1.00000	1.00000	1.00000	0.05872	14.784	2.3444

Table 7-4. Component values for the filter of Figure 7-10 for ω_H = 1 rad/sec.
(*R* in ohms; *C* in farads)
(Reprinted from *Electronics*, November 11, 1976; Copyright © McGraw-Hill, Inc., 1976)

Ripple (dB)	Ripple factor (ϵ^2)
0.01	0.00230524
0.03	0.00693167
0.1	0.023293
0.3	0.0715193
1.0	0.255925

Table 7-5. Ripple factors for equal ripple filters.

(Reprinted from *Electronics*, November 11, 1976; Copyright © McGraw-Hill, v-Inc., 1976)

Example. Design a low-pass equal-ripple filter with a ripple of 0.1 dB and a cutoff frequency of 5 kHz.

Solution. The filter configuration is shown in Figure 7-10. The component values for the normalized filter are obtained from Table 7-4.

$R_1 = R_2 = R_3 = 1$ ohm; $C_1 = 0.09691$ F; $C_2 = 4.7921$ F; $C_3 = 1.3145$ F

To have a cutoff frequency of $\omega_H = 2\pi 5 \times 10^3 = 31.416 \times 10^3$ rad/s, we divide all the capacitors by 31.416×10^3 giving

$C_1 = 3.085 \ \mu F$; $C_2 = 152.537 \ \mu F$; $C_3 = 41.842 \ \mu F$

Finally, to obtain practical component values, we multiply all the resistor values by 10^4 and divide all the capacitor values by the same factor, giving

$R_1 = R_2 = R_3 = 10$ Kohms
$C_1 = 308.5$ pF; $C_2 = 0.0153 \ \mu F$; $C_3 = 0.0048 \ \mu F$

The frequency response of the filter can be obtained with the aid of Equation 7-2. This equation is for the normalized filter. To use it for the denormalized filter, ω must be replaced with f/f_H where f_H is the pass-band cutoff frequency.

$$\frac{V_2}{V_1} = \frac{1}{\left[1 + 0.023293 \left(4 \dfrac{f^3}{1.25 \times 10^{11}} - 3 \dfrac{f}{5 \times 10^3}\right)^2\right]^{1/2}}$$

where we replaced ϵ with 0.023293 as given in Table 7-5 for 0.1 dB ripple. For an illustration, we compute the attenuation at 5,000 Hz and at 7,000 Hz.

At a frequency of 5,000 Hz:

$$\frac{V_2}{V_1} = \frac{1}{[1 + 0.023293\,(4\text{-}3)^2]^{1/2}} = \frac{1}{(1.023293)^{1/2}} = 0.98855$$

In dB

$$20\,\log_{10} 0.9886 = -0.1000 \text{ dB}$$

That is, at 5 kHz the gain is down by 0.1 dB from the d-c gain. At a frequency of 7,000 Hz:

$$\frac{V_2}{V_1} = \frac{1}{\left[1 + 0.023293\,\left(4 \times \dfrac{3.43 \times 10^{11}}{1.25 \times 10^{11}} -3 \times \dfrac{7 \times 10^3}{5 \times 10^3}\right)^2\right]^{1/2}} = 0.6951$$

In dB,

$$20\,\log_{10} 0.6951 = -3.159 \text{ dB}$$

Thus, at 7 kHz the gain is down by 3.159 dB from the d-c gain. Additional points can be computed to plot the complete frequency response curve of the filter.

High-Pass Filter

The circuit configuration for a high-pass filter is shown in Figure 7-11. The component values for the normalized filter are given in Table 7-6. If the cutoff frequency must be $\omega_L = 2\pi f_L$ rather than $\omega = 1$, all the capacitors must be divided by ω_L just as was done in the example for the low-pass filter. Finally, to obtain practical component values, all the resistors are multiplied

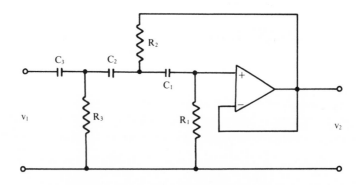

Figure 7-11. Third order high-pass filter.

(Reprinted from *Electronics*, November 11, 1976, copyright © McGraw-Hill, Inc., 1976)

Filter Class	R_1	R_2	R_3	C_1	C_2	C_3
Maximally flat (Butterworth) 3.01 dB at ω_1	4.93949	0.28194	0.71808	1.00000	1.00000	1.00000
Equal ripple (Chebishev) Ripple (dB)						
0.01	10.9130	0.39450	1.18991	1.00000	1.00000	1.00000
0.03	10.09736	0.30186	0.96852	1.00000	1.00000	1.00000
0.10	10.3188	0.20868	0.76075	1.00000	1.00000	1.00000
0.30	11.65230	0.13499	0.59428	1.00000	1.00000	1.00000
1.00	17.0299	0.06764	0.42655	1.00000	1.00000	1.00000

Table 7-6. Component values for the filter of Figure 7-11 for ω_H = 1 rad/sec.
(R in ohms; C in farads)
(Reprinted from *Electronics,* November 11, 1976, copyright © McGraw-Hill, Inc., 1976)

by a given factor and all the capacitors are divided by the same factor. This was also demonstrated for the low-pass filter.

The frequency responses for the normalized low-pass maximally-flat and equal-ripple filters are given respectively by Equations 7-1 and 7-2. For the normalized high-pass filters ω in these equations must be replaced with $1/\omega$. For the denormalization high-pass filters ω in Equations 7-1 and 7-2 must be replaced with f_L/f where f_L is the cut-off frequency.

7.3.4 Fourth-Order Filters

Fourth-order filters are designed by cascading two second-order filters. The circuit diagram of a fourth-order low-pass filter is shown in Figure 7-12 and for a fourth-order high-pass filter in Figure 7-13. The values of the components are determined with the aid of Table 7-7 and Table 7-8, respectively. The tables list component values for a maximally flat filter and for an equal-ripple filter with a ripple of 3 dB. Many different combinations of components yield the right characteristics. In Tables 7-7 and 7-8 we list two possible combinations of components for each stage of each class. Either set of components of the first stage can be combined with either set of

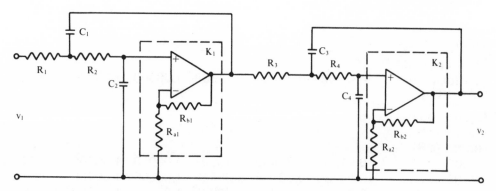

Figure 7-12. Fourth-order low-pass filter.

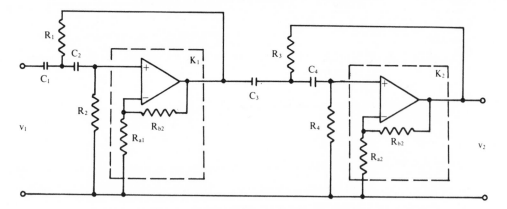

Figure 7-13. Fourth-order high-pass filter.

the second stage of the same class. Components for the first set of each stage are selected so that either both resistors or both capacitors and as many other components as possible are equal to unity. Components for the second set are chosen so that the gain of the amplifier is equal to an integer. This makes it possible to use two resistors of the same value for the amplifier. In particular, a gain of 2 results in resistance values (of R_{a1} and R_{b1} for the first stage and R_{a2} and R_{b2} for the second stage) of unity, since the gain is given by unity plus the ratio of the resistances. In one case (first stage of the low-pass equal-ripple filter), the gain is set equal to 5. A lower integer in this case results in negative values of other components, and is therefore physically unrealizable. We shall illustrate the use of the tables in two examples.

Low-Pass Filter

Example. Design a fourth-order maximally flat filter with a cutoff frequency of 800 Hz.

| Filter Class | Stage | 1 | R_1 | R_2 | C_1 | C_2 | K_1 |
		2	R_3	R_4	C_3	C_4	K_2
Maximally flat	Stage	1	1.0000	1.0000	1.0000	1.0000	1.1522
			1.0000	0.5412	1.0000	1.8478	2.0000
(Butterworth)		2	1.0000	1.0000	1.0000	1.0000	2.2346
			1.0000	1.3065	1.0000	0.7654	2.0000
Equal Ripple	Stage	1	1.0000	5.1020	1.0000	1.0000	5.0041
(Chebishev)			1.0000	5.1225	1.0000	0.4960	5.0000
Ripple: 3 dB	Stage	1	1.0000	1.1073	1.0000	1.0000	2.1986
		2	1.0000	13.6032	1.0000	0.0816	2.0000

Table 7-7. Component values for the filter of Figure 7-12 for ω_H = 1
(R in ohms; C in farads)

| Filter Class | Stage | 1 | R_1 | R_2 | C_1 | C_2 | K_1 |
		2	R_3	R_4	C_3	C_4	K_2
Maximally flat	Stage	1	1.0000	1.0000	1.0000	1.0000	1.522
			1.0000	0.5412	1.0000	1.8478	2.0000
(Butterworth)		2	1.0000	1.0000	1.0000	1.0000	2.2346
			1.0000	1.3065	1.0000	0.7654	2.0000
Equal Ripple		1	1.0000	1.0000	1.0000	0.1960	0.9808
Chebishev	Stage		1.0000	0.1613	1.0000	1.2152	2.0000
Ripple: 3 dB		2	1.0000	1.0000	1.0000	0.9031	2.6358
			1.0000	0.2960	1.0000	0.2673	2.0000

Table 7-8. Component values for the filter of Figure 7-13 for ω_L = 1
(R in ohms; C in farads)

Solution. The circuit diagram is shown in Figure 7-12. From Table 7-7 we choose for the first stage $R_1 = R_2 = 1$ ohm, $C_1 = C_2 = 1F$. The gain of the amplifier is 1.1522. We let $R_{a1} = 1$ ohm, and compute R_{b1}:

$$K = 1 + \frac{R_{b1}}{R_{a1}} = 1.1522 \qquad \frac{R_{b1}}{R_{a1}} = 1.1522 - 1 = 0.1522$$

$$R_{b1} = 0.1522 \, R_{a1} = 0.1522 \text{ ohms}$$

A filter with these components has an upper cutoff frequency $\omega_H = 1$ rad/s.

To shift the cutoff frequency to the specified 800 Hz value, we divide all the capacitors by $2 \pi 800$. This yields

$$C_1 = C_2 = \frac{1}{2\pi 800} = 0.000199 \text{ F}$$

Finally, to obtain reasonable resistor and capacitor values we multiply all resistor values by 10^4 and divide all capacitor values by 10^4.

$$R_1 = R_2 = R_{a1} = 10 \text{ K ohms}; \; R_{b1} = 1.522 \text{ K ohms}$$
$$C_1 = C_2 = 0.0199 \; \mu F$$

Commercially available values are:

$$R_1 = R_2 = R_{a1} = 10 \text{ K ohms}; \; R_{b1} = 1.5 \text{ K ohms}$$
$$C_1 = C_2 = 0.022 \; \mu F$$

The design of the second stage is similar to that of the first stage. From Table 7-7 select

$$R_3 = R_4 = 1 \text{ ohm}; \; C_3 = C_4 = 1F; \; K = 2.2346$$

choosing $R_{a2} = 1$ ohm gives $R_{b2} = 1.2346$ ohms.

Divide the capacitor values by $2 \pi 800$:

$$C_3 = C_4 = 0.000199$$

Multiply all resistors by 10^4 and divide all capacitors by 10^4.

$$R_3 = R_4 = R_{a2} = 10 \text{ K ohm}; \; R_{b2} = 12.346 \text{ K ohm}$$
$$C_3 = C_4 = 0.0199 \; \mu F$$

Commercially available values are:

$$R_3 = R_4 = R_{a2} = 10 \text{ K ohm}; \; R_{b2} = 12 \text{ K ohm}$$
$$C_3 = C_4 = 0.022 \; \mu F$$

The commercially available components are not all equal in value to the computed value. Furthermore, components have tolerances (e.g. resistors ± 5%, capacitors ± 20%). This results in a cutoff frequency different from the specified value. If the specification is critical, the filter must be trimmed using resistor and/or capacitor trimmers.

High-Pass Filter

Example. Design a fourth-order equal-ripple (3 dB) filter with a cutoff frequency of 2 kHz.

Solution. The circuit diagram is shown in Figure 7-13. From Table 7-8 we choose for the first stage

$$R_1 = 1.0000 \text{ ohm}, R_2 = 0.1613 \text{ ohm}$$

$$C_1 = 1.0000 \text{ F}, C_2 = 1.2152 \text{ F}$$
$$K_1 = 2.0000$$

For the second stage we choose

$$R_3 = 1.0000 \text{ ohm}, R_4 = 0.2960 \text{ ohm},$$
$$C_3 = 1.0000 \text{ F}, C_4 = 0.2673 \text{ F}$$
$$K_2 = 2.0000$$

Resistors R_{a1}, R_{b1}, R_{a2}, R_{b2} are computed from $1 + R_{b1}/R_{a1} = K_1 = 2$ and $1 + R_{b2}/R_{a2} = K_2 = 2$. A good choice is $R_{a1} = R_{b1} = R_{a2} = R_{b2} = 1$ ohm. This gives a large number of resistors with an identical value. Divide the capacitors by $2\pi \times 2,000$. This gives

$$C_1 = C_3 = 0.0000796 \text{ F}$$
$$C_2 = 0.0000967 \text{ F}$$
$$C_4 = 0.0000213 \text{ F}$$

Multiply resistor values by 10^5 and divide capacitor values by 10^5.

$$R_1 = R_3 = R_{a1} = R_{b1} = R_{a2} = R_{b2} = 100 \text{ K ohms}$$
$$R_2 = 16.13 \text{ K ohms} \qquad R_4 = 29.60 \text{ K ohms}$$
$$C_1 = C_3 = 796 \text{ pF}, C_2 = 967 \text{ pF}, C_4 = 213 \text{ pF}$$

The designer now chooses appropriate commercially available components and trims the circuit if necessary.

7.3.5   High-Q Band-Pass Filters

The Q of a band-pass filter is defined as the ratio of the center frequency to the bandwidth (i.e. the frequency range between the 3 dB points). A high-Q represents therefore a "sharp" or highly selective filter. The band-pass filters discussed in previous sections are low-Q or broad-band filters (Q of the order of 10). The circuit diagram shown in Figure 7-14 is a high-Q filter (Q's up to several hundreds). It is known as a *state-variable filter*. A design procedure can be formulated by letting, for the normalized filter,

$$R_1 = R_2 = R_3 = R_5 = R_6 = 1 \text{ ohm and}$$
$$C_1 = C_2 = 1F$$

The remaining resistor, R_4, is then computed from

$$R_4 = \frac{2}{\dfrac{1}{Q} - \dfrac{1}{A_1} - \dfrac{1}{A_2}}$$

where A_1 and A_2 are the *open-loop* gains of the amplifiers as shown in Figure 7-14. For gains in the hundreds of thousands, and Q's in the hundreds we have

$$R_4 \simeq 2Q$$

The open-loop gain A_3 must be high, but does not enter the calculation. The gain at resonance ($\omega_c = 1$ rad/sec for the normalized filter) is R_4/R_1. Since R_1 is 1 ohm, the gain is numerically equal to R_4.

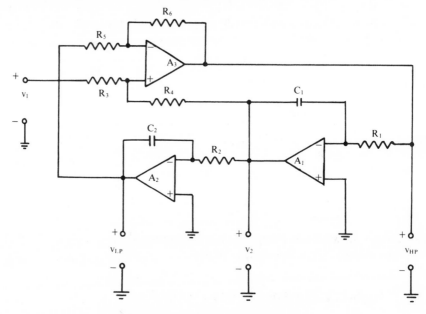

Figure 7-14. State variable filter.

We shall illustrate the use of these equations and show how to design for a desired resonance filter with practical component values in a design example.

Example. Design a high-Q band-pass filter with a center, or resonance, frequency at 5 kHz and a Q of 250.

Solution. We select amplifiers with gains of 10,000, resistors (except R_4) of 1 ohm, and capacitors of 1F. The normalized filter is obtained by setting

$$R_4 = \frac{2}{\dfrac{1}{250} - \dfrac{1}{10,000} - \dfrac{1}{10,000}} = 526.314 \ \Omega$$

To shift the center frequency to $f_c = 5,000$ Hz we divide the capacitors by $\omega_c = 2\pi 5,000$,

$$C_1 = C_2 = \frac{1}{2\pi 5,000} = 63.662 \ \mu F$$

Finally, to obtain practical values for all the components we multiply all the resistors by 10^3 and divide all the capacitors by 10^3. The resistor and capacitor values become:

$$R_1 = R_2 = R_3 = R_5 = R_6 = 1 K\Omega$$
$$R_4 = 526.316 \ K \ \Omega$$
$$C_1 = C_2 = 0.06366 \ \mu F$$

The designer now specifies commercially available components and incorporates capacitor, and/or resistor trimmers if necessary.

Note: The state variable amplifier can also be used as a low-pass filter and as a high-pass filter if the proper output terminals are used as shown in Figure 7-14. This is an expensive version in comparison with other low-pass and high-pass filters, but has the advantage of low Q-sensitivity to circuit parameter variations. The d-c gain and cutoff frequencies for the low-pass filter are

$$\text{Gain} = \frac{1 + R_5/R_6}{1 + R_3/R_4}$$

and

$$f_H = \frac{1}{2\pi} \left(\frac{R_6}{R_1 R_2 R_5 C_1 C_2} \right)^{1/2}$$

The high-frequency gain and cutoff frequency for the high-pass filter are

$$\text{Gain} = \frac{1 + R_6/R_5}{1 + R_3/R_4}$$

and

$$f_L = \frac{1}{2\pi}\left(\frac{R_4}{R_1 R_2 R_5 C_1 C_2}\right)^{1/2}$$

These expressions simplify for the case $R_1 = R_2 = R_3 = R_5 = R_6$ and $C_1 = C_2$ which we used for the band-pass filter.

7.4 HOW AMPLIFIER PERFORMANCE AFFECTS FILTER PERFORMANCE

The non-ideal characteristics of the operational amplifier such as offset voltage, bias current and frequency dependency of the gain affect filter performance and must be taken into consideration, as discussed in Chapter 1, if specifications are very tight. The most critical parameters are perhaps the bias current and gain effects. As to the bias current it is necessary to have the currents into the inverting and the noninverting terminals flow through the same effective resistance values so that the same voltage drop exists at both terminals. A d-c analysis must be performed to trace the paths of the bias currents. In the circuit of Figure 7-7, for example, a resistor $R_3 = R_1 + R_2 - R_a R_b / R_a + R_b$ must be inserted between the junction of R_a and R_b and the inverting input as shown in Figure 7-15.

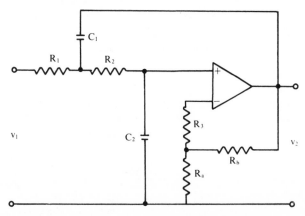

Figure 7-15. Bias current compensation for the filter of Figure 7-7.

With this resistor in place, the total resistance between the inverting terminal and ground is

$$R_3 \quad + \quad \frac{R_a R_b}{R_a + R_b}$$

or

$$R_1 + R_2 \quad \frac{R_a R_b}{R_a + R_b} \quad + \quad \frac{R_a R_b}{R_a + R_b} + R_1 + R_2$$

which is equal to the resistance seen between the noninverting input terminal and ground. Note that in Figure 7-7 the effective resistance between the inverting terminal and ground is the parallel combination of R_a and R_b

$$R_a \| R_b \quad = \quad \frac{R_a R_b}{R_a + R_b}$$

The frequency dependency of the gain of an operational amplifier must be considered. The gain decreases beyond a certain frequency. In filters which use voltage followers (e.g., Figure 7-10) a compensation network can be incorporated as shown in Figure 7-16. As the voltage frequency increases,

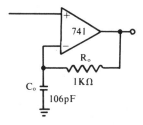

Figure 7-16. Frequency compensation to shunt feedback current at high frequency.

more of the current fed back from the output through R_o flows through C_o and less into the amplifier, thus decreasing the negative feedback and maintaining higher gain. The component values are obtained from the relation

$$R_o C_o \quad = \quad \frac{1}{2.6 \times 2\pi f_o}$$

where f_o is the 0 dB frequency of the amplifier. For the 741 amplifier, for example, f_o = 581 kHz giving a value $R_o C_o = 105 \times 10^{-9}$.

The Q of the high-Q band-pass filter of Section 7.3.5 is limited by the amplifier frequency response. To have a filter essentially independent of the break frequencies of the amplifiers, the amplifier must be selected so that

$$Q \ll A_o f_p / 4 f_c$$

where f_p is the 3 dB break frequency of the amplifier, and f_c is the resonance frequency of the filter.

7.5   FILTER SENSITIVITY

Since the filter characteristic depends on the values of the components and the gain of the amplifiers, it changes as these parameters drift with temperature and time. The highest sensitivity is usually due to changes of amplifier gain. The state variable filter has relatively low sensitivity. In practice, the sensitivity to various parameter changes must be obtained experimentally in the laboratory, and operational amplifiers which result in filters which meet the characteristics must be selected.

7.6   FILTER TUNING

The various tables and design equations give values of much higher precision than can be obtained in commercial fixed value components. If exact frequency and/or gain values are required it is necessary to use resistance and/or capacitance trimmers. In some cases there is interaction between various characteristics. In such cases, several adjustment iterations are necessary.

7.7   PITFALLS TO AVOID

In the design procedures presented in this chapter it was assumed that the operational amplifier has a flat frequency response. To have the designs valid it is necessary that the break frequency of the amplifier be higher than the highest frequency of interest. Otherwise, frequency compensation must be employed such as discussed for example in Section 7.4. If the design specification is tight, bias current effect can become significant depending on the amplifier and on the resistance values. If necessary, bias compensation must be employed as discussed in Section 7.4.

The computed resistance and capacitance values are usually not of standard commercial products. When the filter is built with standard available components, trimmers must be incorporated in the design. Remember also that capacitor tolerances are on the order of +80%, –20%, or ±20% and resistor tolerances ±10% or ±5% which would necessitate trimmers anyway. Don't overlook thick film and thin film resistors and capacitors if your product involves a large number of filters. Film components can be trimmed to order during the fabrication process and can result in savings.

Whether bulk or thin film components are used, they are all subject to drifts with temperature. The filter must be tested over the required temperature range and adjusted or compensated as needed.

Chapter **8**

Basic Digital I.C.'s

8.1   The advent of low-cost MSI (medium-scale digital I.C.'s) and LSI (large-scale digital I.C.'s) makes it possible to design quite complex logic systems with relatively few I.C. packages. An LSI memory, for example, can be the equivalent of 1024 flip-flop gates. A cost effective design aims at minimizing the package count (rather than the individual gate count which was the aim just a few years ago). In general, a good design should make use of LSI and MSI packages as much as possible, and use SSI (small-scale digital I.C.'s) such as NAND and NOR gates only when necessary to fill in the "gaps", or in very small sub-systems.

There is a need for some SSI gates in most systems. Three technologies have become most widely accepted for use in digital I.C.'s: TTL, CMOS, and ECL. Each technology has its unique merits. The device application implications of these technologies are described in this chapter in conjunction with SSI digital circuits; the same design implications are also true for MSI and LSI applications. In Section 8.2 we introduce the basic logic functions and the *truth-table* concept. In Section 8.3 we discuss the truth-table further, and in Section 8.4 we introduce flip-flop gates. TTL, CMOS, and ECL technologies are discussed in Section 8.5. The use of LSI and MSI I.C.'s is presented in Chapter 9.

8.2   LOGIC FUNCTIONS AND THE TRUTH TABLE

In digital systems, variables (voltages) assume only two distinguishable levels: (a) *high level*, and (b) *low level*. A voltage applied to a device, or put out by a device, is recognized as *high* if it is above a specified value. The voltage is *low* if it is below a specified value. The specified values depend on the technological family of which a particular device is a member

(TTL, CMOS, or ECL). Independent of the particular device, the two levels are symbolized by a "1" and a "0". This notation is called *positive logic notation*. In *negative logic* the high level voltage is designated as "0" and the low level voltage as "1". By far the more common descriptions in the literature are in terms of positive logic. We shall use in this book positive logic unless specifically indicated otherwise. In fact, all designs can be carried out thinking in terms of positive logic, even if the designer encounters a so-called negative logic device in the system. This point will be elaborated further in Section 8.2.8.

There are six basic logic functions: AND, OR, NOT, NAND, NOR, and EXCLUSIVE OR. We define these functions and introduce the corresponding symbols and truth tables in the next few paragraphs. The several logic functions specify under what input conditions there is an output, or there is no output. The *truth table* is a tabular representation of this information.

8.2.1   The AND Function

A device which performs the AND function has a "1" output if, and only if, *all* the input terminals are "high," that is, if all the input terminals receive a "1" input. In terms of the symbolic representation of Figure 8-1a, F is "high" if, and only if, A and B and C are "high". This information is presented very effectively in a tabular form as shown in Figure 8-1b, called the "Truth Table." This table shows, for example, that if input A is "low," input B is "low," and input C is "high," then output F is "low" (second row). If input A is "low," input B is "high," and input C is "high," then output F is "low" (fourth row). If input A is "high," input B is "high," and input C is "high," then output F is "high" (eighth row).

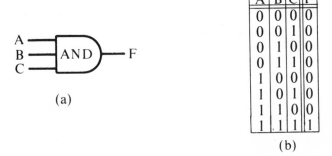

(a)

A	B	C	F
0	0	0	0
0	0	1	0
0	1	0	0
0	1	1	0
1	0	0	0
1	0	1	0
1	1	0	0
1	1	1	1

(b)

Figure 8-1. AND Function (shown for three input terminals).
(a) Symbol
(b) Truth table

8.2.2   The OR Function

A device which performs the OR function has a "1" output if any one or any combination of the input terminals is "high." In terms of the symbolic representation of Figure 8-2a, F is "high" if inputs A or B or C or any combination of these is "high." This information is conveyed in the truth table given in Figure 8-2b. Note that the existence of a "high" output does not require that only one input is "high," but that *at least* one input is "high."

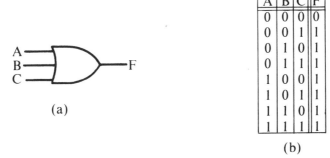

A	B	C	F
0	0	0	0
0	0	1	1
0	1	0	1
0	1	1	1
1	0	0	1
1	0	1	1
1	1	0	1
1	1	1	1

(a)

(b)

Figure 8-2. OR Function (shown for three input terminals).
(a) Symbol
(b) Truth table

8.2.3   The EXCLUSIVE OR Function

A device which performs the EXCLUSIVE OR function has a "1" output if one, and only one, input is "high." In terms of Figure 8-3a the output is "high" if A or B or C is "high." The output is "0" if no input, or more than one input, is "high." The truth table for the device is given in Figure 8-3b.

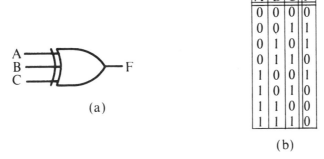

A	B	C	F
0	0	0	0
0	0	1	1
0	1	0	1
0	1	1	0
1	0	0	1
1	0	1	0
1	1	0	0
1	1	1	0

(a)

(b)

Figure 8-3. EXCLUSIVE OR Function (shown for three input terminals)
(a) Symbol
(b) Truth table

The *two input* EXCLUSIVE OR device can also be used to perform other functions: (a) it can be used to compare two signals. There will be a "high" output only if the two inputs are *not* the same, i.e., only if A ≠ B; (b) it can be used as a *controlled inverter*. If one of the input terminals is considered to be a control input terminal, then if the control is "high" the input to the other terminal is inverted. If the control is "low," then the input the other terminal is not inverted.

8.2.4   The INVERTER or NOT Function

The INVERTER inverts the input signal, or performs the NOT function. The symbol and truth table are given in Figure 8-4.

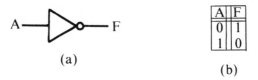

(a)

A	F
0	1
1	0

(b)

Figure 8-4. The INVERTER or NOT Function
(a) Symbol
(b) Truth table

8.2.5   The NAND Function

NAND stands for NOT AND. That is, the output is complementary to the AND output. It is analogous to the output of an AND device followed by an INVERTER. Thus, the output is "0" if, and only if, inputs A and B and C in Figure 8-5a are "high." The truth table for the NAND device is given in Figure 8-5b.

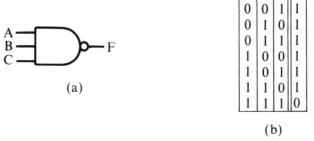

(a)

A	B	C	F
0	0	0	1
0	0	1	1
0	1	0	1
0	1	1	1
1	0	0	1
1	0	1	1
1	1	0	1
1	1	1	0

(b)

Figure 8-5. The NAND Function (shown for three input terminals).
(a) Symbol
(b) Truth table

8.2.6   The NOR Function

NOR stands for NOT OR. That is, the output is complementary to the OR output. It is analogous to the output of an OR device followed by an INVERTER. Thus, the output is "0" if input A or input B or input C in Figure 8-6a, or any combination of these inputs, is "high." The truth table of the NOR device is given in Figure 8-6b.

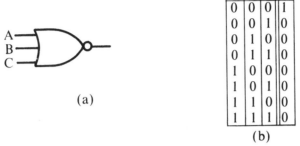

A	B	C	F
0	0	0	1
0	0	1	0
0	1	0	0
0	1	1	0
1	0	0	0
1	0	1	0
1	1	0	0
1	1	1	0

(a)

(b)

Figure 8-6. The NOR Function (shown for three input terminals).

(a) Symbol
(b) Truth table

8.2.7   Rise Time and Delay Time

In the design of logic circuits the required input and output signals are pulses. An ideal pulse is shown in Figure 8-7a; the rates of change of voltage levels are infinite. Real pulses have finite rates of change, and finite *rise times* and *fall times* as shown in Figure 8-7b. The rise time and fall time are indicated in the figure as t_{dr} and t_{df} respectively. *Edge-triggered* devices

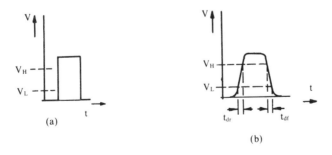

(a)

(b)

Figure 8-7. Binary pulses.

(a) ideal pulse
(b) real pulse

require a rate of change higher than a certain minimum for the triggering to occur. For level-triggered signals, the voltage levels must be equal to v_H or higher, or equal to v_L or lower for triggering to occur. The specifications for given devices are supplied by the manufacturer.

There is always a delay time between the instant of signal application to an input terminal of a device and the instant of the appearance of an output signal in response to the input signal since the transmission of signals and switching requires finite time. The delay times must be considered in the design of switching systems. Consider, for example, the logic circuit shown in Figure 8-8. This circuit has an output (i.e. T "high") if A is "low" and B is "high" or if A is "high" and B is "low." If A and B change states simultaneously, the new state of B will exist at the input of gate 1 before the new state of A is transmitted through an inverter. To have gate 1 respond correctly, the new values of A and B must be present at the input to gate 1 simultaneously. The designer must take the delay into consideration and make certain that both new signals are present at gate 1 at the time that the output of gate 1 is important.

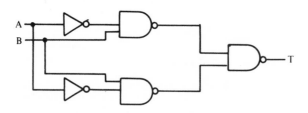

Figure 8-8. Logic circuit.

8.2.8   Positive True and Negative True Logic

Positive and *negative* logic are also called *high true* and *low true* logic. In positive (or high true) logic a high voltage (e.g. 5 V) means logic "1" while a low voltage (e.g. 0 V) means the logic "0". In negative (or low true) logic a low voltage means a logic "1", and a high voltage means the logic "0". As a result of these conventions a NAND gate used in a positive logic circuit behaves as a NOR gate in a negative logic circuit. Consider a logic gate represented by a square block in Figure 8-9a with a truth table given in Figure 8-9b. In this figure L represents *low* voltage and H represents *high* voltage.

For *positive* logic the truth table is as given in Figure 8-10a. Comparing this table with Figure 8-5b, we see that we have a NAND gate the symbol of which is given in Figure 8-5a (except that in the present

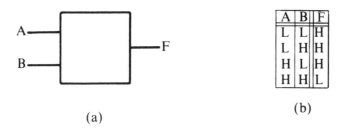

(a)

A	B	F
L	L	H
L	H	H
H	L	H
H	H	L

(b)

Figure 8-9. General logic circuit.

(a) Block diagram
(b) Truth table

A	B	F
0	0	1
0	1	1
1	0	1
1	1	0

(a)

A	B	F
1	1	0
1	0	0
0	1	0
0	0	1

(b)

Figure 8-10. Truth tables.

(a) NAND circuit
(b) NOR circuit

example we have only two inputs A and B). For *negative* logic the truth table of Figure 8-9b takes the form of the table in Figure 8-10b. Comparing this table with Figure 8-6b, we find that we have a NOR gate (for two inputs). The symbol for this negative logic circuit is shown in Figure 8-11.

Figure 8-11. Negative logic NOR circuit.

Note that the small circles (bubbles) can be considered to represent *negative* or *low true* logic input terminals. The absence of small circles represents *positive* or *high true* logic. Accordingly the symbol in Figure 8-11 can be interpreted as a *low true* input, *high true* output OR circuit, and therefore as a low true (negative logic) NOR circuit. The symbol in Figure 8-5a can be interpreted as a *high true* input, low true output AND circuit and therefore a high true (or positive logic) NAND circuit.

These concepts and symbols are used in some of the literature and the reader should be familiar with them. They appear confusing. Fortunately in practical work it is not necessary to think in complicated terms; it is only important to know whether at a given point in the logic circuit there will be a high voltage or whether there will be no high voltage under a set of input conditions. This information can be obtained by thinking only in terms of positive logic AND and OR circuits replacing the small circles at the input and the output of gates with inverters. The logic identities are shown in Figure 8-12. Be careful when considering the delay times to take into account only the delay of the single gate; not the delay of the equivalent gates. It should be noted that small circles are also used to indicate *negative-edge* triggering in more complicated circuits. For example a shift register may shift one position during the *positive-going edge* (positive rate of change of voltage) of a pulse when applied to a given terminal, or during the *negative-going edge* when applied to another terminal. The latter terminal is identified with a small circle.

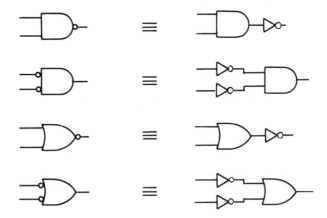

Figure 8-12. Logic symbol identities.

8.3 TRUTH TABLE

We introduced the concept of a truth table in Section 8.2. The truth table simply states what logic output ("1" or "0") will exist for different combinations of inputs. We have seen truth tables for the various basic logic functions. More complex I.C.'s also have corresponding truth tables which help the designer to apply the I.C.'s. Unless otherwise specified, positive logic is assumed. That is, a "1" represents high voltage and a "0" represents low voltage.

As an example, Figure 8-13 gives the block diagram of a *Decimal Decoder*. The inputs are coded in the binary number system and the outputs

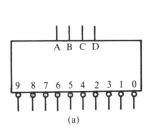

(a)

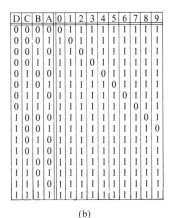

D	C	B	A	0	1	2	3	4	5	6	7	8	9
0	0	0	0	0	1	1	1	1	1	1	1	1	1
0	0	0	1	1	0	1	1	1	1	1	1	1	1
0	0	1	0	1	1	0	1	1	1	1	1	1	1
0	0	1	1	1	1	1	0	1	1	1	1	1	1
0	1	0	0	1	1	1	1	0	1	1	1	1	1
0	1	0	1	1	1	1	1	1	0	1	1	1	1
0	1	1	0	1	1	1	1	1	1	0	1	1	1
0	1	1	1	1	1	1	1	1	1	1	0	1	1
1	0	0	0	1	1	1	1	1	1	1	1	0	1
1	0	0	1	1	1	1	1	1	1	1	1	1	0
1	0	1	0	1	1	1	1	1	1	1	1	1	1
1	0	1	1	1	1	1	1	1	1	1	1	1	1
1	1	0	0	1	1	1	1	1	1	1	1	1	1
1	1	0	1	1	1	1	1	1	1	1	1	1	1
1	1	1	0	1	1	1	1	1	1	1	1	1	1
1	1	1	1	1	1	1	1	1	1	1	1	1	1

(b)

Figure 8-13. Decimal decoder

(a) Block diagram

(b) Truth table

correspond to decimal numbers. All this information is summarized in the truth table (Figure 8-13b). For example, for A = 1, B = 0, C = 0, D = 0 which represents a binary "one," all the outputs except terminal 1 are *high*. The low output at terminal 1 causes a light marked "1" to go on. For A = O, B = 1, C = 1, D = 0, which represents a binary "six", all outputs, except terminal 6, are high.

8.4   LATCH AND FLIP-FLOP GATES

A latch gate responds to an input and locks (latches) the output into this state until it is instructed otherwise. The output of a flip-flop gate changes state (from "0" to "1" and vice-versa) each time an input is applied to the gate. Latch and flip-flop gates are discussed following the introduction of asynchronous and synchronous operation.

8.4.1   Asynchronous and Synchronous Operation

In asynchronous operation the output responds immediately (except for a small delay time) to a change in the input signal. In synchronous operation the output does not respond immediately to the input. It requires the presence or occurrence of another input, the *clock pulse,* to affect the change of state at the output. This is important in large systems since by proper timing it is possible to ensure that all the inputs are present where needed when clocking and change of state occur. This eliminates false triggering resulting from delays, capacitive and inductive coupling, and noise.

The clock input signal can be one of three types: (a) d-c or edge-triggered, (b) a-c coupled, or (c) master-slave. The d-c or edge-triggered device changes state when a particular voltage level is reached. Some gates respond to the positive edge of the pulse, and others to the negative edge of the pulse. The time rates of rise and fall of the pulses are not critical. An a-c triggered device responds to a rate of change of voltage; the clock is capacitively coupled internally to the latching mechanism of the flip-flop. The rise and fall times of the pulse are critical and must be typically less than 200 ns. The master-slave operation applies to a flip-flop which consists basically of two latches connected in series, the first being the "master" and the second the "slave." The action of the clocking pulse consists of four steps (Figure 8-14). At instant t_1 the slave is isolated from the master; at instant t_2 the input is enabled to accept new data; at instant t_3 the input is disabled, and at instant t_4 the data is transferred from master to slave. In these devices the data inputs are never connected directly to the outputs. This provides total isolation of the outputs from data inputs.

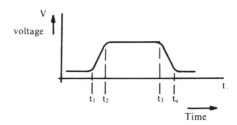

Figure 8-14. Master-slave clock pulse.

8.4.2 The R-S Latch

The block diagram and the truth tables of the R-S latch are shown in Figure 8-15. There are two truth tables, depending on the internal structure

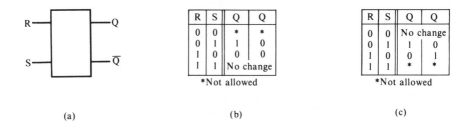

(a)   (b)   (c)

Figure 8-15. R-S latch.

(a) Block diagram
(b) Truth table for R-S NAND latch
(c) Truth table for R-S NOR latch

of the latch. The inputs are designated R (reset or clear) and S (set or preset). The outputs are designated Q and $\overline{Q}$, each being the complement of the other. In one structure, input R = 0, S = 0 is ambiguous and is not allowed. In the other structure input R = 1, S = 1 is ambiguous and is not allowed.

8.4.3   The D-Latch

The block diagram of the D-latch (Data latch) is shown in Figure 8-16. The clock pulse acts in this case as an "enable" input. When the clock is high, the input at D appears directly at the output Q. During this time any changes in the D input are immediately reflected at the Q output. When the clock falls, the Q output holds the state of the D input prior to the negative-going clock edge regardless of the input until the clock goes high again. The *preset* and *clear* input, when present, override the D and clock inputs: if R and S are *high*, they are inactive; if R is *low*, Q is *high*, and if S is *low*, Q is *high* regardless of D and *clock* inputs. R and S may not both be low.

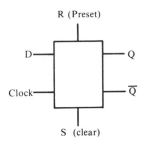

Figure 8-16. D-type latch.

8.4.4   The D-type Flip-Flop

The symbol is the same as that shown in Figure 8-16. The truth table is shown in Figure 8-17. The instant t_n is just prior to clocking and the instant t_{n+1} is just after clocking. *Preset* and *clear* inputs may also be present. They override the D and *clock* inputs as was described in Section 8.4.3.

t_n	t_{n+1}
D	Q
0	0
1	1

Figure 8-17. Truth table of D-type flip-flop.

8.4.5   The J-K Flip-Flop

This is the most widely used flip-flop. The symbol and truth table of the J-K flip flow are shown in Figure 8-18. The figure also shows the *excitation table*. This table shows what *inputs are needed* to obtain a *desired output* at t_n+_1, if the output at t_n is known. It is derived from the truth table. The J-K flip-flop may also have *preset* and *clear* input terminals. Inputs into these terminals override other input signals as was described in Section 8.4.3. Note that when J and K are held high, then the output changes state each time a clock pulse arrives.

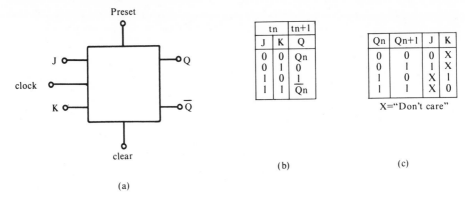

	tn		tn+1
	J	K	Q
	0	0	Qn
	0	1	0
	1	0	1
	1	1	$\overline{Qn}$

Qn	Qn+1	J	K
0	0	0	X
0	1	1	X
1	0	X	1
1	1	X	0

X="Don't care"

(a)

(b)

(c)

Figure 8-18. J-K flip-flop.

(a) Symbol
(b) Truth table
(c) Excitation table

8.4.6   Commercially Available Gate Packages

There is a large variety of the various flip-flop types and other gates available in many different packages. For example, there are flip-flops which are triggered by the positive edge of the clock pulse, and others which are triggered by the negative edge. There are individually packaged flip-flops, and there are packages which contain two flip-flops. Similarly, there are single-packaged NAND Gates and multiple-packaged NAND Gates. The Hex-Inverter is very popular. This is a package which contains six separate inverters. The obvious advantage of the use of multiple gate packages is that space is conserved, only one set of power supply connections is needed and the total cost is lower.

There are several different technologies available for the manufacture of gates which perform the same logic functions. Each technology has its own advantages and disadvantages in terms of the resulting performance characteristics such as speed and power handling capability. The various technologies are discussed in Section 8.5 below.

8.5   DIGITAL CIRCUIT TECHNOLOGIES

Of the various technologies which have evolved over the years, three have reached and maintained a position of prominence: TTL logic (transistor-transistor-logic), CMOS logic (complementary-metal-oxide-semiconductor logic), and ECL logic (emitter-coupled logic).

8.5.1   TTL Technology

TTL I.C. logic elements have been the most widely used for many years. They implement a compromise between speed and power consumption. The most widely used are the *Schottky* and the *Low Power Schottky* families. The Schottky devices have a switching time of 3 nsec/gate and power dissipation of 20 mW/gate. The Low Power Schottky devices have a switching time of 5 nsec/gate and power dissipation of 2mW/gate. Earlier families which are to be found in existing systems, but which are rarely employed in new designs, are: *Standard* (switching time of 10 nsec/gate and power dissipation of 10 mW/gate); *Low power* (33 nsec/gate and 1 mW/gate); and *High speed* (6 nsec/gate applicable to clock frequencies of 50 MHz, 23 mW dissipation). The designations of the five families are in the order listed above 54/74S, 54/74LS, 54/74, 54/74L, and 54/74H. The postscripts designate the families while 54 and 74 designate the series.

There are two series of TTL devices: the 54 series and the 74 series. The nominal specifications for all the gates are: supply voltage 5.0 V; logical "0" output 0.2 V; logical "I" output 3.0 V; noise immunity 1.0 V. The 54 series meets the military specifications which call for a supply voltage tolerance of 4.5 V to 5.5 V over a temperature range from 55°C to 125°C. The 74 series meets the commercial specifications which call for a supply voltage tolerance of 4.75 V to 5.25 V over a temperature range from 0°C to 70°C.

It is important to recognize that when an input into a gate is "0" then the gate which sets the level "0" must *sink* a current into its own output terminal. The total current that it can sink determines the *fan-out* of the gate, i.e., the number of gates that can be connected to its output terminal.

Open Collector Gates

At times it is desirable to perform a logic function which can be described by the following example: We want a "0" output if A and B, or C and D and E, or F and G are "1". In principle this logic function can be performed by the circuit of Figure 8-19. Such a circuit should not be used for the following reason. Suppose one of the NAND gates has a "0" output while the other two have "1" outputs. The current which must be "sinked" by the NAND with "0" output will be excessive since the current includes also currents from the other two NAND gates. This will not only degrade the "0" level, but can destroy the NAND gate with a zero output.

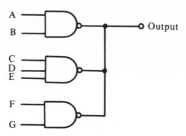

Figure 8-19. This circuit must not be used (see text). In principle it performs the function: no output if there is an input at *A and B, or C and D and E, or F and G.*

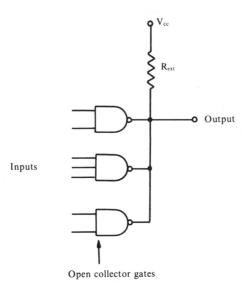

Figure 8-20. Wired-AND logic (wired-OR logic). Three open collector NAND gates wired together.

To implement this function, one uses open collector gates as shown in figure 8-20. This procedure is sometimes called wired-OR logic. The pull-

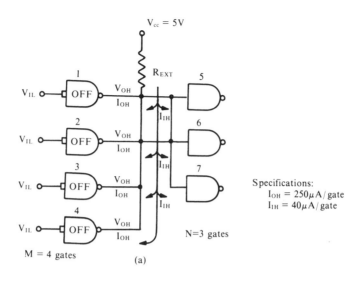

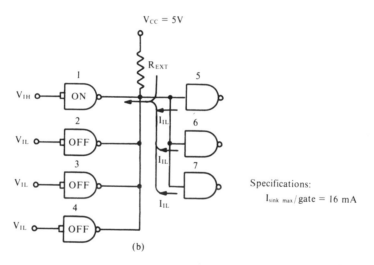

Figure 8-21. Diagrams for computation of R_{ext}.
(a) Condition for computation of R_{ext} max.
(b) Condition for computation of R_{ext} min.

up resistor R_{ext} is supplied externally by the designer. There is one resistor common to all the gates. The value of R_{ext} must be computed. A maximum limit and a minimum limit are established as is demonstrated in the following example with the aid of Figure 8-21. The double subscripts IL, IH, OL, and OH designate respectively "input low," "input high," "output low," and "output high." The maximum permissible value of R_{ext} is determined for the condition of all low input voltages (Figure 8-21a). The outputs of gates 1 through 4 are high; the currents flowing into the output terminals of the gates are the leakage currents I_{OH}. The inputs into gates 5 through 7 are high and the current flowing into each gate is I_{IH}. For a "high" output voltage we require a minimum value of 2.4 V. This gives the maximum allowable voltage drop across R_{ext} as $V_{cc} - V_{OH\ Req'd}$.

Thus,

$$(MI_{OH} + NI_{IH})\ R_{ext}\ max = V_{cc} - V_{OH\ Req'd}$$

or

$$R_{ext}\ max = \frac{V_{cc} - V_{OH\ Req'd}}{MI_{OH} + NI_{SH}} =$$

$$\frac{5V - 2.4V}{4 \times 0.250\ mA + 3 \times 0.040\ mA} =$$

$$\frac{2.6V}{0.00112\ A} = 2321\ ohms.$$

The minimum value of R_{ext} is obtained when one of gates 1 through 4 is ON and the others are OFF as shown in Figure 8-21b. The maximum output voltage V_{OL} must not exceed a certain value, say 0.4 V. The ON gate must now *sink* the currents coming from gates 5 through 7 plus the current flowing through R_{ext}. The minimum value of R_{ext} is therefore obtained as

$$R_{ext}\ min = \frac{V_{cc} - V_{OH\ Req'd}}{I_{sink\ capability} - I_{sink\ from\ TTL\ loads}} =$$

$$\frac{5.0V - 0.4V}{16mA - 3 \times 1.6\ mA} = \frac{4.6V}{0.0112A} = 410\ ohms$$

We thus have the design value

$$410\ \Omega < R_{ext} < 2321\ \Omega$$

A 1.5 K ohm resistor is suitable.

Three-State Logic (Common Bussing)

Another method of connecting outputs of gates together without overloading any gate, independent of the states of the other gates, is accomplished through the use of "three state" gates. The output of such a gate can be a "one," a "zero" or an *open circuit*.

If one of the outputs is "zero" and it is desired to disconnect the other gates from the bus, the outputs are set to the "open circuit" state in which case they appear as high impedances across the bus line.

Noise Immunity

It is important to have well-defined output voltages for "0" and "1" levels. The output corresponding to a "0" output is guaranteed to be between 0 and 0.4 V and the output corresponding to a "1" output is guaranteed to be between 2.4 V and V_{cc}. So that the input of a stage will recognize the "0" and "1" outputs of a previous stage even at the presence of noise which may raise the voltage of a "0" above 0.4 V or lower the output of a "1" below 2.4 V, the gates are guaranteed to recognize a "0" if the voltage is between 0 and 0.8 V and a "1" if the voltage is between 2.0 V and V_{cc}. These limits are guaranteed for worst V_{cc} conditions over the entire guaranteed operating temperature range. The guaranteed noise margin is thus 0.4 V for both logic states. Typically, noise immunity is over wider margins, in excess of 1.0 V. At 25° C gates typically change states as the input voltage passes through 1.35 V. The output is typically 3.3 V in the logical "1" state and 0.2 V in the logical "0" state. Therefore, in the logical "1" state the output can typically tolerate 1.5 V of negative-going noise on the line connecting the two gates before causing the driven gate to change state falsely. Similarly, 1.15 V of positive-going noise can typically be tolerated on a line in the logic "0" state. The 0.4 V margin is, however, guaranteed under all specified permissible operating conditions. In designs for mass production, guaranteed values, rather than typical values, must be considered.

The *duration* of a noise signal also enters the picture. The switching times of TTL gates are in the nanosecond range. A noise pulse of a microsecond duration is in comparison "long"; it is referred to as a d-c noise. The noise margins discussed above were for d-c noise in this sense. Shorter noise pulses (referred to as a-c noise) can have higher magnitudes before they cause false switching. (The reason is that a certain amount of charge transfer must occur before switching occurs.) For example, at 25° C and V_{cc} = 5 V, the noise immunity to pulses of duration of 9 ns or longer is nearly 2 V, and for shorter pulses the immunity is greater than 2 V increasing indefinitely for pulses of less than 6 ns duration.

8.5.2   CMOS Technology

CMOS or COS/MOS (or other similar terms) stand for "Complementary symmetry Metal-Oxide-Semiconductor". These gates are fabricated of combinations of *n*-channel and *p*-channel MOS transistors, which gives them their unique characteristics. CMOS gates constitute the

"4000 logic gate series." Two important features of CMOS gates are their extremely low quiescent power dissipation and their ability to operate under a wide range of supply voltage. The switching times of the gates decrease with increasing supply voltage, but are longer than the switching times of TTL gates. The recommended supply voltage range (i.e. voltage between drain and source) is between 3 V and 15 V. The switching times depend on the particular device, loading, and temperature, but typical values at 25° C and a load of 15 pF are 22 ns for 15 V, 25 ns for 10 V, 35 ns for 5 V, and 50 ns for 3 V for SSI devices and increase for MSI and LSI devices to several hundred nanoseconds. Typical quiescent power dissipation values are 5 nW for 5 V supply and 10 nW for 10 V supply for SSI gates, increasing to a few μW for LSI devices.

A CMOS gate normally drives other CMOS gates. Since the inputs to these devices are insulated gates, they constitute purely capacitive loads. The loads are typically between 5 pF and 15 pF/gate. Except during switching, no currents flow through the output terminals. At times, CMOS devices are interconnected with other gate families such as TTL gates. In this case the CMOS gate may have to supply current to the next stage (in which case it acts as a current source) when the CMOS gate output is high, or it may have to sink a current when its output is low. Typical output source and sink current capabilities are a few, or a few tens, milliamperes.

Noise Immunity

The guaranteed noise immunity for most devices is ± 3 V for 10 V operation and ± 1.5 V for 5 V operation. Thus, for example, if a device operates with a supply voltage of 10 V and the input is a logic "1" (i.e. 10 V), the device will not change state when the input level drops from 10 V to 7 V because of noise. If the input is a logic "0" (i.e. 0 V), the device will not change state when noise signals of 3 V appear at the input. *Typical* noise immunity is better than the *guaranteed* noise immunity. Typical noise immunities are ± 4.5 V for 10 V operation and 2.25 V for 5 V operation. Some devices have different guaranteed noise immunities as indicated on the data sheets. In design for mass production, guaranteed, rather than typical, values must be considered.

Three-State Logic (Common Bussing)

CMOS cannot be wire-ORed like TTL gates because of the complementary nature of the basic inverter circuit. But CMOS gates can be common-bussed by the use of three-state transmission gates which are incorporated in the NAND, NOR or other gates. Under this arrangement the output of a gate can be a "1", a "0" or an *open circuit*. More than 50 such gates can be connected to one bus. If one of the outputs is "0" and it is desirable to disconnect the other gates from the bus, their outputs are set to the "open circuit" state.

Comments

All the inputs of a gate must be connected to V_{DD} (the most positive potential), to V_{SS} (the most negative potential), or to a signal source. A floating input terminal makes the output unstable and unpredictable. Whether to connect an unused input terminal to V_{DD} or V_{SS} or together with another input terminal to a signal source depends on the device and the desired output under specified conditions. For example, the output of a NAND gate remains "high" unless all the inputs are "high." Therefore, an unused input must be connected to V_{DD} so that when all the control gates are "high" the output will switch to "low." For analogous reasons the unused terminals of a NOR gate must be connected to V_{SS}.

Gates on the same chip can be paralleled as shown for a NOR gate in Figure 8-22. This increases the source and sink output current capabilities, and the increased drive capability increases the speed. The total power dissipation is, of course, also increased.

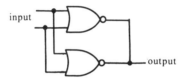

Figure 8-22. Device paralleling (NOR gates serve only as example).

8.5.3   ECL Technology

The *E*mitter *C*oupled *L*ogic is particularly suitable for high-speed operation. The ECL 10,000 family has gate delay propagation times of 1.5 ns (10,200 and 10,600 series) and of 2 ns (10,100 and 10,500 series), flip-flop toggle speeds of 200 MHz and 125 MHz, and gate powers of 25 mW. The gates specified for industrial use (ECL 10,100 and ECL 10,200) are guaranteed to operate in an ambient temperature range from $-30°C$ to $+85°C$. The military units (ECL 10,500 and ECL 10,600) are specified for operation in a temperature range from $-55°C$ to $+125°C$. The high-speed operation is possible because the devices do not operate in their saturation regions and thus are not slowed down by junction charge storage.

Another very important characteristic of ECL gates is that they cause a nearly *constant* power supply current drain, without spikes during the signal transition period. This "quiet" operation is a result of the differential amplifier circuitry of the devices. As the output switches, the current through one transistor decreases while the current through another transistor increases by the same amount. The differential amplifier configuration of the ECL gates makes it easy to make an output and its complementary signal available in the same device. Consequently, even the simple NAND and NOR gates have complementary outputs available. Thus a NOR gate, for example, can also be used as an OR gate (Figure 8-23).

Figure 8-23. An OR-NOR gate symbol.

The difference between the high and the low voltage of the required power supply for ECL gates is 5.2 V. Best noise immunity is achieved when the high voltage V_{CC} is grounded ($V_{CC} = 0$ V) and the low voltage is set at V_{EE} = –5.2 V. Other values of V_{EE} may be used. A more negative voltage will increase the noise margin, but it will also cause more power dissipation. A less negative voltage will have the opposite effect. The "high" state is –0.9 V and the low state is –1.75 V, providing an output logic swing of 0.85V. For positive logic a logic "1" corresponds to –0.9 V and a logic "0" corresponds to –1.75 V.

Noise Immunity

The noise margin, i.e., the maximum noise which will not cause false switching, is 0.125 V for switching from "high" to "low" and 0.155 V for switching from "low" to "high." The maximum output voltage that a gate may have when it is in the "low" state is –1.630 V. The maximum allowable input voltage gate which does not cause a transition from "low" to "high" is –1.475 V. Thus, a positive spike of up to –1.475 V – (–1.630 V) = 155 mV will not cause switching. The corresponding values for "high" to "low" switching are –1.105 V and –0.980 V. Noise spikes of duration shorter than the gate delay time can be higher without causing false transitions.

Unused Inputs

Unused inputs must be connected to a negative voltage to ensure that no voltage across the high input impedance of the differential amplifier builds up which could cause false triggering. Unless specified otherwise for a particular device by the manufacturer, V_{EE} can serve as the negative voltage.

Comments

Because of the high speed of operation certain factors must be considered which can be ignored in the application of slower devices. The interconnecting wiring causes time delays in transmitting the signal. Every foot of interconnecting wiring introduces a delay of two nanoseconds, or about the equivalent of one gate. This delay time can be reduced only by reducing the length of the interconnecting lines. Wave distortion due to line reflection is also of concern. At the high speeds, line length can approach the wavelength of the signal and improperly terminated lines can result in reflections that cause false triggering. This can be eliminated by terminating each signal line with its characteristic impedance. This is accomplished by connecting a resistor from the end of the line (usually before it connects to the receiving gate) to a negative voltage. A voltage of -2 V can be used, but to avoid the need for an additional voltage, the resistor can be returned to V_{EE} of -5.2 V. The higher voltage magnitude causes higher dissipation. Common resistor values are 150, 100, 75, 50 ohms when returned to -2.0 V, and 2.0 K, 1.0 K, 680, 510, and 270 ohms when returned to -5.2 V. When terminating resistors are used the power dissipated by the resistor must be added to the total power dissipation and the additional power dissipation of the output transistor which supplies the resistor must be added to the gate package dissipation. Table 8-1 lists power dissipation values for commonly used resistors.

Terminating Resistor Value	Output Transistor Power Dissipation (mW)	Terminating Resistor Power (mW)
150 ohms to −2.0 Vdc	5.0	4.3
100 ohms to −2.0 Vdc	7.5	6.5
75 ohms to −2.0 Vdc	10	8.7
50 ohms to −2.0 Vdc	15	13
2.0 K ohms to V_{EE}	2.5	7.7
1.0 K ohms to V_{EE}	4.9	15.4
680 ohms to V_{EE}	7.2	22.6
510 ohms to V_{EE}	9.7	30.2
270 ohms to V_{EE}	18.3	57.2
82 ohms to V_{CC} and 130 ohms to V_{EE}	15	140

Table 8-1.* Typical power dissipation in output circuit with external terminating resistors.

*From MECL Integrated Circuit Data Book, Motorola Semiconductor Products, Inc.

8.5.4   Mixing Logic Families

In some designs it may be advantageous to use different logic families in the same system. For example, high speed devices may be necessary in one part of the system while advantage of low power devices can be taken in other parts of the system. Gates of different families can be interconnected directly or by means of interface translators. Proper interfacing between different logic families requires that the gates display logic level compatibility including sufficient noise margins. Current input requirements and current output capabilities must also be considered and fan-out capabilities determined accordingly.

TTL/CMOS Interfacing

TTL gates can drive CMOS gates directly, but to ensure operation and enough noise margin for a logic "1" output from the TTL, the use of a pull-up resistor R_p as shown in Figure 8-24 is advisable. If the TTL 7400 series is used, V_{DD} must not be above 5.25 V and not below 4.75 V. For the TTL 54 series the limits are 5.5 V and 4.5 V. The input current to a CMOS gate is 10 pA. For the "low" or "0" logic state, the output of the TTL at 10 pA is guaranteed to equal or be less than 0.4 V. The CMOS will recognize an input as "0" as long as the input is below 1.5 V, giving a noise margin of 1.1 V. The no load "high" or "1" TTL output can be as high as 3.6 V, but under load the output is only guaranteed to have a minimum voltage of 2.4 V. When the pull-up resistor R_p is used, the "1" output is equal to V_{DD} ensuring proper operation with adequate noise margin. The "0" output voltage will also be increased and a proper choice of R_p must be made depending on the fan-out.

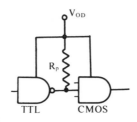

Figure 8-24. TTL gate driving CMOS gate.

TTL gates can be driven directly by certain CMOS gates. For a logic "1" the CMOS output is at least 4.4 V, and the required TTL input is at most 2 V. For a "0" input of a TTL the input must not be greater than 0.4 V and the required current sinking capability of the CMOS must be 1.6 mA per gate. This capability is offered by the CMOS 4009 and CMOS 4010 gates at an "0" output voltage of 0.4 V. Other CMOS gates can be used if several input terminals of the CMOS gate are paralleled as this increases the current sinking capabilities. A few examples are shown in Figure 8-25. The 4009 A and 4010 A gates can also be used as level shifters. They can accept a high voltage, such as 15 V, at one terminal supplying one part of the circuit and 5 V at a second terminal which makes the interfacing with the TTL gate possible.

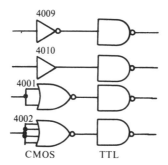

Figure 8-25. CMOS gates driving TTL gates.

CMOS/ECL Interfacing

As already noted, ECL gates work best when connected to a power supply 0 and –5.2 V. CMOS can operate with the same power supply. In fact, a power supply of –5 ± 1V is satisfactory for the combination of gate

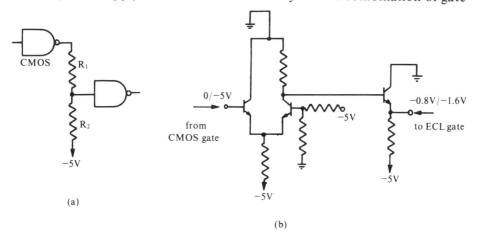

Figure 8-26. CMOS to ECL level translators.
(a) Passive voltage divider circuit.
(b) Active circuit.

families. However, logic level translators are necessary since the logic levels are otherwise not compatible.

Consider a CMOS gate driving an ECL gate. For logic "1" the nominal output voltage of the CMOS gate is 0 V, and the nominal input is –0.8 V. For logic "0" the respective values are –1.6V and –4.2V. Appropriate level shifting circuits are shown in Figure 8-26. A level translator to connect an ECL gate to a CMOS gate is shown in Figure 8-27.

TTL/ECL Interfacing

Commercial I.C.'s are available for interfacing TTL and ECL gates. Examples are the level translators 10124 and 10125. Their delay times are about 5 ns. They require a 5 V power supply (also used for the TTL gates) and a –5.2 V power supply (also used for the ECL gates). If it is desired to use only one power supply, discrete-component translators can be built to bring the logic levels into compatibility.

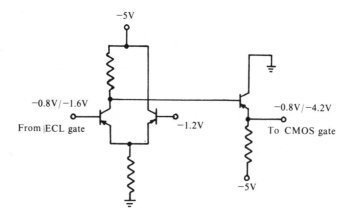

Figure 8-27. ECL to CMOS level translator.

Chapter 9

Digital Building Blocks

9.1   Improvements in semiconductor processing methods and various technical breakthroughs have allowed manufacturers to produce increasingly complex digital integrated circuits. The late sixties saw the introduction of single chip I.C.'s with circuit complexities equivalent to 20 to 100 simple gates, so called Medium Scale Integration (MSI). Large Scale Integration (LSI) soon followed with greater than 100 gate complexities, providing the circuit designer with powerful digital building blocks. MSI and LSI circuits offer a reduced package count when replacing Small Scale Integrated (SSI) circuits in a given system. Further benefits include increased system speed and reduced power consumption since a particular complex function (for example, a 4-bit synchronous counter) can be realized in integrated form using fewer and/or simpler gates than its counterpart constructed with SSI gates and flip-flops.

Using MSI/LSI, the designer's perspective is elevated allowing him to view the overall system problem unhampered by the details of each building block. In this chapter, we shall discuss in some detail available MSI/LSI building block functions and their applications. System architecture considerations and binary number systems and their arithmetics are also treated.

9.2   DIGITAL CODING

All information in a digital system is coded in digital form (1's and 0's). In simple systems this information may have limited meaning; representing, for example, the state of some switch (off or on) or the desired state of some other switch. In general, however, more complex systems must deal with both control states *and* data in numerical form. There are various means of representing a number in digital form and, depending on the particular application some representations are more useful than others. By far the most common coding scheme is the natural binary system.

9.2.1   Natural Binary Coding

In the straight binary scheme a number N is represented as a weighted sum of increasing powers of two, i.e.,

$$\text{MSB} \qquad\qquad\qquad\qquad \text{LSB}$$

$$N = a_n \times 2^{n-1} + a_{n-1} \times 2^{n-1} + \ldots + a_2 \times 2^1 + a_1 \times 2^0 \qquad \textbf{(9-1)}$$

The weighting coeffecients $(a_n \ldots a_1)$ represent the binary coding of the number N. For example, substituting the 7-bit binary number 1001101 in Equation 9-1 we have, since n = 7,

$$
\begin{array}{ccccccc}
a_7 & a_6 & a_5 & a_4 & a_3 & a_2 & a_1 \\
\end{array}
$$

$$\text{Binary No.} = \quad 1 \quad\; 0 \quad\; 0 \quad\; 1 \quad\; 1 \quad\; 0 \quad\; 1$$

$$N = 1 \times 64 + 0 \times 32 + 0 \times 16 + 1 \times 8 + 1 \times 4 + 0 \times 2 + 1 \times 1$$

$$N = 77$$

The largest number that can be represented by an N-bit binary number is $2^{n+1} - 1$.

To convert a decimal number to binary form, perform a series of divisions by 2, keeping track of the remainder. For example, to convert 77 to binary form we have,

	Quotient		Remainder
77/2	=	38	1
38/2	=	19	0
19/2	=	9	1
9/2	=	4	1
4/2	=	2	0
2/2	=	1	0
1/2	=	0	1
Thus,	77	=	1 0 0 1 1 0 1

9.2.2   **Binary Arithmetic**

Binary arithmetic operations are simpler to perform than their decimal equivalents

Addition

The truth table for binary addition is shown in Table 9-1. For example, let A = 1010110 (86) and B = 1110101 (117), find A + B.

		Binary			Decimal
carry	→	1 1 1 1	carry		11
A	=	1 0 1 0 1 1 0			86
B	=	1 1 1 0 1 0 1			117
A + B	=	1 1 0 0 1 0 1 1			203

The addition proceeds from right to left according to the truth table.

A + B	carry in	result	carry out
0 + 0	0	0	0
1 + 0	0	1	0
0 + 1	0	1	0
1 + 1	0	0	1
0 + 0	1	1	0
1 + 0	1	0	1
0 + 1	1	0	1
1 + 1	1	1	1

Table 9-1. Truth table for binary addition.

Subtraction

The subtraction A (minuend) – B (subtrahend) can be performed using the truth table for digital addition by properly coding the subtrahend. In a digital system this has the advantage of using the adder stages to perform both addition and subtraction. The coding scheme is quite simple. Consider the decimal subtraction 8 – 4 = 4. By taking the tens complement of the subtrahend i.e. (10–4 = 6) and adding this to the minuend (8) we have

$$\begin{array}{r} 8 \\ +6 \\ \hline \text{carry} \rightarrow \quad 1)4 \end{array}$$ and if we ignore the carry,

we generate the correct result, 4. Unfortunately, in decimal notation this process still requires a subtraction to find the 10's complement, but in binary notation the equivalent of 10's complement, i.e. the 2's complement, can be formed without using subtraction. To form the 2's complement, proceed as follows. Step 1 — replace all 1's with 0's and all 0's with 1's (this step is called the one's complement) then, Step 2, add "1" to the result of Step 1. As an example of digital subtraction using 2's complement consider the subtraction 8-4:

We write   1000 (8) minuend
               −0100 (4) subtrahend

Step 1: Form 1's complement of subtrahend so 0100 becomes 1011 then
Step 2: Add one to the result thus 1011 + 1 = 1100

Then add the minuend and the 2's complemented subtrahend and ignore the carry:

                    1000 (8) minuend
     Step 3:   +1100 two's complemented subtrahend
              carry→ 1)0100 (4) answer

The sign of the carry indicates the polarity of the difference: 1 = (+) and 0 = (−). If the subtrahend is larger than the minuend, then the difference is negative and the carry is 0. *Note, however, that when the difference is negative, the answer is coded in 2's complement notation.*

Multiplication.

Procedurally, binary multiplication is identical to decimal multiplication. As an example, consider the multiplication 10101 (21) × 10110 (22) we have;

Since the multiplier consists of only 1's and 0's, each partial product is either zero or the multiplicand itself. As in decimal multiplication, the rightmost bit of each partial product is positioned under the bit in the multiplier which generated that particular partial product. Finally, the partial products are added to obtain the result.

Division

Binary division is best explained by an example. The procedure is identical to decimal division.

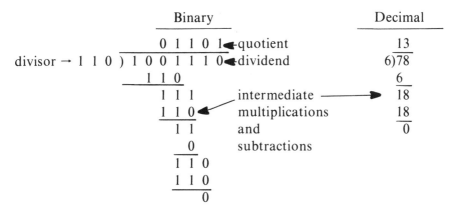

The preceding binary arithmetic is valid for natural binary coded numbers only. It is usually best to convert a number that is represented in another coding scheme back to natural binary to perform arithmetic operations.

Note that all binary arithmetic operations can be performed using an adder, an inverter (used to generate the one's complement) and some control signals. For binary subtraction, we need only generate the 2's complement of the subtrahend and proceed with the addition. In the case of multiplication the problem is the correct positioning of the partial products (which are either zero or identical to the multiplicand) and then we simply add them up. Similarly, with division we must position the intermediate multiplications (which are either zero or identical to the diviser) and perform the indicated subtractions which become additions using the 2's complement technique.

9.3   BINARY CODED DECIMAL (BCD)

The BCD code is formed by converting each digit in the decimal number into a 4-bit natural binary number and placing the 4-bit number in the position occupied by the original digit in the decimal number. For example,

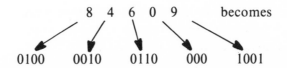

A space is usually left between the 4-bit numbers to aid in reading the number back in decimal form. A 4-bit number can represent a maximum of 16 numbers, but in the BCD coding scheme only 10 of these numbers are used and, therefore, the BCD scheme is an inefficient way to store numbers in digital form. For example, in natural binary 84609 can be coded using only 16 bits instead of the 20 bits required for BCD coding. Because the conversion from decimal to BCD and vice versa is so simple, this code finds greatest use at the inputs and outputs of user-oriented digital systems. BCD arithmetic is possible, but very complicated as compared to natural binary arithmetic.

9.4 USING MSI/LSI INTEGRATED CIRCUITS

The available MSI/LSI circuits fall into several classes or groups according to their function. We will treat each class separately and discuss the use of the MSI/LSI circuits which are included in that class.

9.4.1 Adders

As we demonstrated in Section 9.2.1 a binary adder must be used to perform any of the arithmetic operations. It is, therefore, a very important digital sub-system. Looking at Table 9-1 we see that a simple adder stage will have three inputs: A, B and carry-in from the previous less significant stage, and two outputs: the result and a carry-out to the next more significant

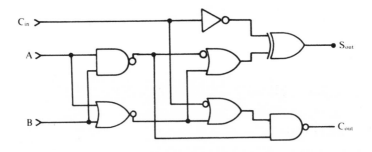

Figure 9-1. Single full adder stage.

stage. Figure 9-1 is a realization of the truth table in hardware form. This basic stage can be connected to add two n-bit numbers in parallel as shown for n = 3 in Figure 9-2. This connection is commonly called ripple carry since each more significant stage must wait to receive valid carry-out data from the preceding less significant stages before producing a valid result. If it takes, say, 120 ns for each stage to generate a valid carry-out, then the final outputs of an n-bit adder will not be valid until n × 120 ns after the binary numbers are applied to the inputs. Thus, for a 16-bit parallel input ripple-carry adder 16 × 120 ns = 1.920 μs would be required for each addition. This speed limitation can be greatly minimized through the use of the "look ahead" carry-out generator described in the next section.

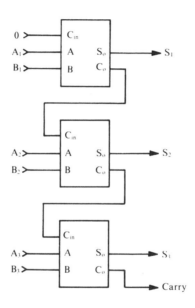

Figure 9-2. Three bit parallel adder with ripple carry.

4-Bit Full Adder with Look Ahead Carry-Out

A block diagram of a 4-bit adder which uses a look ahead carry-out scheme is shown in Figure 9-3. As before, adder stages 1-4 are connected with ripple carry-outs. However, a carry generator has been added. This circuit generates the carry-out that would have come from adder No. 4, but with fewer gate delays. Recall that if a carry-out occurs from a 4-bit adder, this means that the result of the addition is larger than 15. The look ahead carry-out generator simply looks at the inputs in parallel and determines if the sum will be larger than 15. The circuit of Figure 9-3(b) is a detail of the MC14008, an MSI CMOS I.C., which is functionally identical to Figure 9-

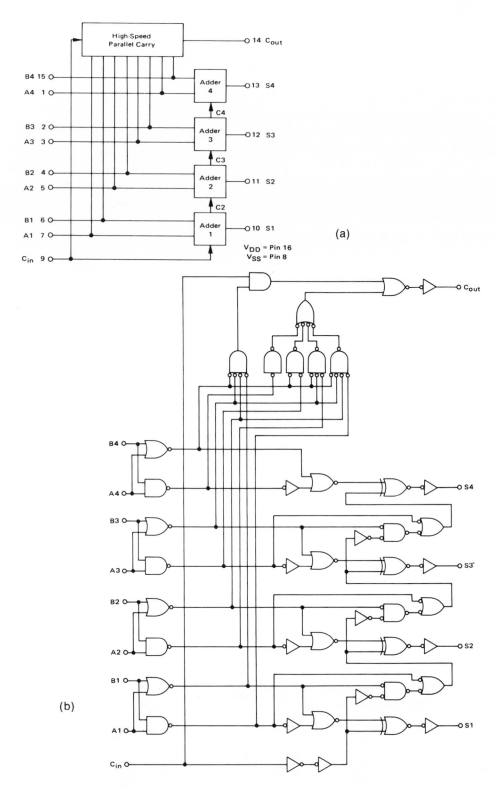

Figure 9-3. 4-bit adder with look-ahead carry outputs. (Courtesy of Motorola Semiconductor Products, Inc.)

288

3(a). Using the look ahead scheme, the total gate delay from carry-in to carry-out across 4 bits is reduced from approximately 14 gate delays to 3 gate delays.

The speed increase realized with this technique is very evident when large bit numbers are added. Consider the 16-bit parallel adder of Figure 9-4. With a delay between carry-in and carry-out of 120 ns for each 4-bit adder, the final carry-out will be valid in 480 ns. This represents a factor of 4 increase in operating speed. Using a low power Schottky TTL MSI circuit such as the LS283 4-bit adder with fast carry, add time would be reduced to about 40 ns.

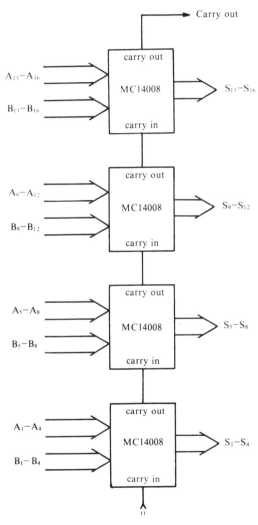

Figure 9-4. 16-bit adder using four 4-bit adders with look-ahead carry output.

If speed is not important, then a large savings in parts count can be realized by using a serial adder. With a single full adder and a D-type flip-flop, a bit serial adder can be constructed as shown in Figure 9-5. Operation begins with the clearing of the flip-flop which sets carry-in to zero. The least significant bits (A_1, B_1) of the two numbers to be added are applied to the A and B inputs generating S_1 out and the proper carry information at C_o. With a positive clock transition the carry-out is latched by the flip-flop presenting the correct carry-in for the adder to proceed with the addition of A_2 and B_2. The flip-flop acts as a one-bit memory storing the carry-out from a previous full addition to be used as the carry-in for the next full addition. Clearly, any word length can be added using this technique. Since large numbers of parallel lines need not be routed around a printed circuit board, bit-serial addition yields a further savings in board area.

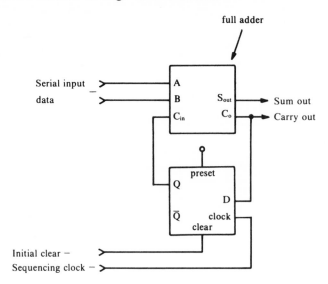

Figure 9-5. Bit serial adder.

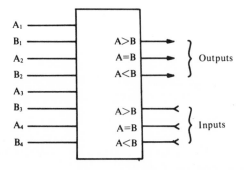

Figure 9-6. 4-bit magnitude comparator the 54/74 LS85.

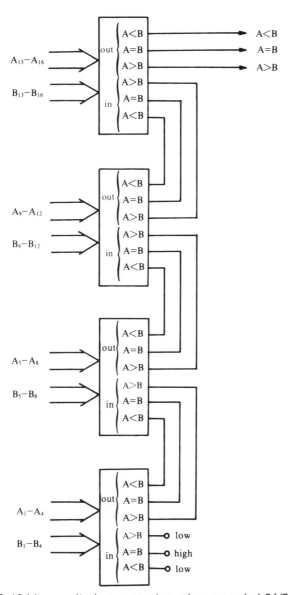

Figure 9-7. 16-bit magnitude comparator using cascaded 54/74 LS85's.

9.4.2   Comparators

A digital comparator performs the function of determining whether a binary number A is less than, equal to, or greater than another binary number B. A functional diagram of the 54/74 LS85 4-bit magnitude comparator is shown in Figure 9-6. In addition to the 8-bit input, three outputs are provided — A> B, A = B, A< B, along with three cascading inputs A>B, A = B, A<B. For a simple 4-bit comparison, set A = B input high and set A>B, A<B low. The active high output will then indicate the relative magnitudes of the two 4-bit numbers A and B applied to the input. By using the cascading inputs a comparison of any word length can be performed as shown in Figure 9-7. Long word length magnitude comparators are especially useful in numerical machine controls. Figure 9-8 shows a simplified linear position control system. By storing a series of desired positions the machine could be automatically stepped through a sequence of positions.

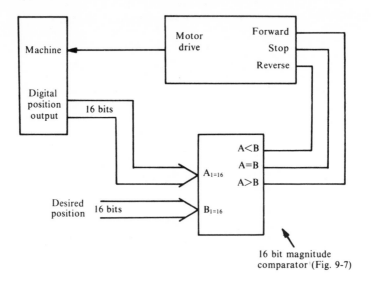

Figure 9-8. Linear position control.

9.4.3 Parity Generators and Checkers

In complex systems digital data is frequently transferred from one location to another. There is always the chance that the number which arrives at the receiving end differs from the number which was sent because of an error somewhere along the transmission path. If noise or a "soft" device failure causes the error, then retransmission of the data could result in a correct transmission. The problem is determining *at the receiving end* if the data has survived the transmission process unaltered. To do this, additional data must be transmitted along with the original data and this supplemental information must tell something about the character of the original data.

A fairly effective error detection technique is to transmit along with the data word an additional bit called a *parity bit*. The circuit of Figure 9-9 is

TRUTH TABLE

W	data	Q	data + parity bit
0	odd	1	even
0	even	0	even
1	odd	0	odd
1	even	1	odd

NOTE: Odd & even refer to the numbers
of 1's in that particular word.

LOGIC DIAGRAM

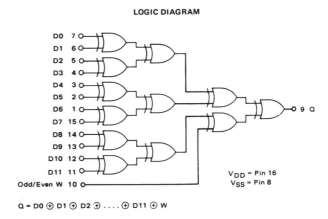

$Q = D0 \oplus D1 \oplus D2 \oplus \ldots \oplus D11 \oplus W$

Figure 9-9. Parity generator/checker, MC14531. (Courtesy of Motorola Semiconductor Products, Inc.) Truth Table

used to generate this parity bit. If odd parity is desired, the MC14531 examines the 12 input lines and choses the parity bit so that the total number of ones in the transmitted word, (including the parity bit) is odd. For even parity, the parity bit is chosen so that the total number of ones (including the parity bit) is even. For most applications odd parity is chosen so that at least one "1" is transmitted in any word. At the receiving end another MC14531 is used with the 13 bits applied in any order to the 12D lines and the W line. If odd parity is received, the Q output will be high and vice versa for negative parity. If the incorrect parity is received the receiving end will request a retransmission of that particular data word.

Adding the parity bit will only indicate a single bit error with no indication of the particular bit at fault. If an even number of bits are in error, the parity will be unchanged and the error will go unnoticed. But in practice, addition of a parity bit to a data transmission system has proved to be a cost-effective method of limiting transmission errors in digital systems.

9.4.4   Shift Registers

A shift register consists of a connection of parallel clocked flip-flops as shown in Figure 9-10. When the clock line is static, the data stored in stage n is presented to the D input of stage n + 1 and the input of stage n is seeing the data stored in stage n – 1. At the positive transition of the clock line, data at the D input of each flip-flop is transferred to its Q output. Thus, each bit of data *shifts* one position to the right. For proper circuit operation, the data hold time requirement at the D input must be less than the propagation delay of the previous stage. Fortunately, this condition is guaranteed for all D-type flip-flops produced today.

The use of MSI/LSI technology has allowed manufacturers to integrate several flip-flops on a single chip thus providing many shift stages in a single package. The simplest n-stage shift registers have one serial input and one serial output separated by n shift states. However, several variations are possible as discussed below.

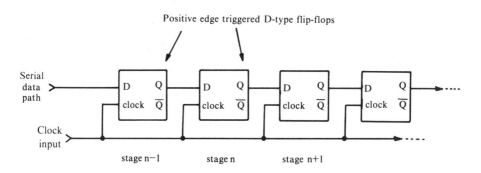

Figure 9-10. Basic shift register.

Serial Input-Parallel Output

A simplified functional diagram of the 54/74LS164 serial input-parallel output 8-bit shift register is shown in Figure 9-11. A clear input is provided to allow asynchronous clearing of each stage and the Q output of each flip-flop is brought out. This circuit can be used as a serial-input, serial-output shift register with from 1 to 8 shift stages. More generally, however, this type of register is used as a demultiplexer converting multiplexed serial data into parallel data.

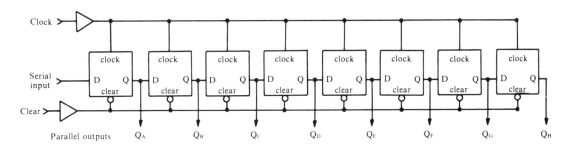

Figure 9-11. 8 bit serial input parallel output shift register.

Parallel Input-Serial Output

This type of shift register has parallel inputs to each stage and a serial input to the first flip-flop. A parallel-serial control line determines which set of inputs is active. The CMOS shift register shown in Figure 9-12 also has parallel outputs from the last three stages. The 4014 has a synchronous parallel load and the 4021 an asynchronous parallel load. When the serial input is selected, serial data is entered on the positive transition of the clock. When the parallel mode is selected on the 4014, parallel data is entered on the positive transition of the clock (synchronous load). With the 4021, however, parallel data is jammed into the flip-flops when the parallel/serial line is taken high, independent of the clock state (asynchronous load). This type of shift register finds primary application in multiplexing parallel data on to a single serial output line.

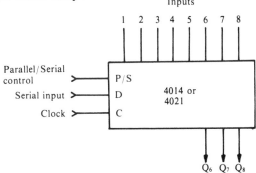

Figure 9-12. Parallel input serial output shift registers CD4014 or CD4021.

9.4.5   Counters

Digital counters are perhaps the most widely used MSI/LSI circuits. Applications range from simply keeping track of time to the control of complex digital systems. As a result, many types of counters are available, each with its particular area of applicability. To simplify our discussion, we shall begin by dividing the general class of counters into those which operate asynchronously (ripple clock) and those which operate synchronously (parallel clocked).

Asynchronous Counters

Figure 9-13 shows a simple asynchronous counter constructed of negative-edge triggered J-K flip-flops. With the J-K inputs tied high, each flip-flop will change state at the negative-going transition of its clock. The term asynchronous is applied to this counter since the output transition of each flip-flop does not occur simultaneously. Consider time t_1; the input clock goes negative and because of the propagation delay of f-f No. 1, Q_1 goes low a few nanoseconds later. The negative transition of Q_1 sets Q_2 low after a few nanoseconds delay. Finally, Q_2 sets Q_3 low, again, a few nanoseconds later. Thus at cardinal transition points such as t_1 and t_2 the clocking signal must *ripple* through all the previous stages before reaching the most significant flip-flop. Unfortunately, between the time the first flip-

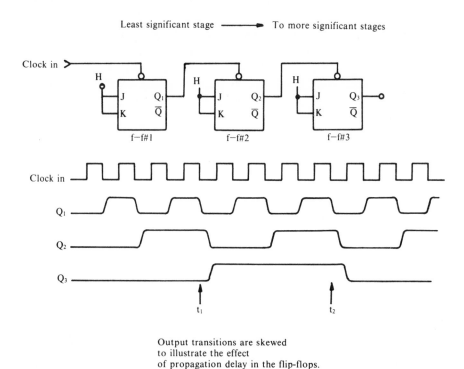

Output transitions are skewed
to illustrate the effect
of propagation delay in the flip-flops.

Figure 9-13. Asynchronous binary counter and timing diagram.

flop is clocked and the final flip-flop settles to its correct value, the digital code on the output line switches through many incorrect codes. In simple counting or dividing applications these incorrect codes cause no problems, but in more sophisticated applications some problems arise as discussed below.

Synchronous Counters

Many counters are used as control elements in sequential systems. As such, the counter outputs (usually binary coded) are applied to a logic bank which decodes predetermined states and provides the control signals to the proper locations at the correct time. Using an asynchronous counter, a high-speed decoding bank would sense the incorrect codes that can occur, as described above, causing errors. In a synchronous counter, however, all outputs switch to their correct states simultaneously since the flip-flops are clocked in parallel. To understand how this is accomplished, consider the circuit diagram of Figure 9-14. With the J-K inputs of f-f No. 1 tied high its output changes state at each positive clock transition. Flip-flop No. 2 will change state if the J-K inputs are held high prior to the clock transition, but this occurs only when Q_1 is high. Therefore, f-f No. 2 divides the frequency of f-f No. 1 in half (as was the case with the asynchronous counter), but because f-f No. 1 and 2 are clocked simultaneously, the output transitions also occur simultaneously. Achieving synchronous operation in a many-stage counter requires considerable circuitry as the circuit diagrams of Figure 9-15 demonstrate.

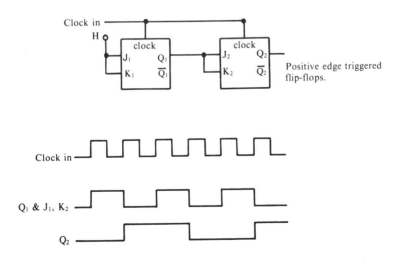

Figure 9-14. 2-stage synchronous counter and timing diagram.

LOGIC DIAGRAMS

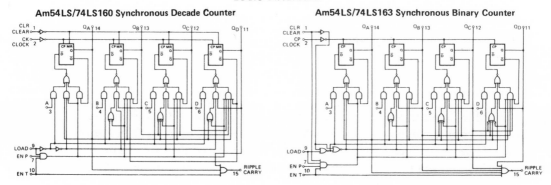

Am54LS/74LS160 Synchronous Decade Counter   Am54LS/74LS163 Synchronous Binary Counter

Figure 9-15. Synchronous counters. (Copyright © 1978 Advanced Micro Devices, Inc. Reproduced with permission of copyright owner.)

Modulo N Counting

The circuits of Figure 9-15 also have preset inputs which control the states of the outputs if the *load* line is low. Operation is synchronous and when *load* is low, data on the preset inputs is transferred to the outputs on the positive clock transition. Using this feature, the LS161 can be made to divide by any number from 1-16. Figure 9-16 shows a typical modulo N counter. When a count of 15 is reached, *ripple carry* goes high setting *load* low. On the next positive clock transition the number on the preset line, in this example 12, is loaded into the output. With each positive clock transition a normal binary counting sequence continues from 12 until 15 is reached and the sequence repeats. As shown in the accompanying table any division modulo can be achieved by a proper selection of the preset number.

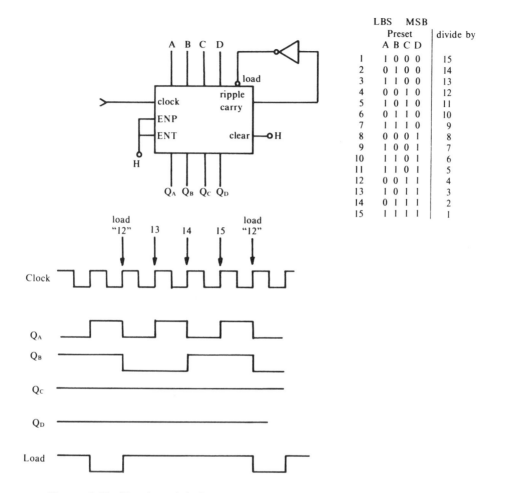

	LBS MSB Preset A B C D	divide by
1	1 0 0 0	15
2	0 1 0 0	14
3	1 1 0 0	13
4	0 0 1 0	12
5	1 0 1 0	11
6	0 1 1 0	10
7	1 1 1 0	9
8	0 0 0 1	8
9	1 0 0 1	7
10	1 1 0 1	6
11	1 1 0 1	5
12	0 0 1 1	4
13	1 0 1 1	3
14	0 1 1 1	2
15	1 1 1 1	1

Figure 9-16. Circuit and timing diagrams for "modulo N" counting.

Chapter 10

Analog to Digital Conversion

10.1   Hybrid analog/digital electronic circuits are becoming increasingly common in complex systems. Typical examples of such systems include:

- Data acquisition and reduction systems where analog data is digitised and operated on by a microprocessor or other digital processor.

- An analog memory system where the incoming analog data is digitised and stored in random access memory for later retrieval. Such a system is cheaper and faster than magnetic tape in many cases.

- Cathode ray tube display, such as numeric displays on oscilloscope screens. In this case, the desired number or image is stored in digital form and applied to a digital analog converter which drives the deflection plates to reproduce the image on the tube face.

Many other such systems could be described, but one requirement is common to all, that of converting analog voltages to digital numbers and vice versa. Thus, an analog to digital (A/D) converter and/or a digital to analog (D/A) converter must be included in these systems. As a class these converters are frequently referred to as A/D/A converters since A/D and D/A circuits have many similarities.

Thus the hybrid systems designer is faced with the task of selecting an A/D/A converter type and designing it into his system. This job is rarely simple since many A/D/A converter design approaches exist and each has its special advantages and disadvantages. In this chapter, we will present the basic considerations common to all converters e.g. resolution, coding formats, etc. and proceed to discuss in depth the most generally used conversion systems and appropriate design data.

10.2   ADA CONVERTER BASICS

The purpose of an ADA converter is to associate an analog voltage such as 118 volts with a digital number such as 01110110. The most common digital coding scheme is natural binary with the MSB (Most Significant Bit) listed first. In this code a binary number of n digits is related to its decimal equivalent N by

$$N = a_n^{MSB}\, 2^{n-1} + a_{n-1}\, 2^{n-2} + \ldots a_3 2^2 + a_2 2^1 + a_1^{LSB} \times 1 \qquad \textbf{(10-1)}$$

where a_n through a_o represent the binary digits in the n bit number. Calculating the decimal equivalent of 01110110 we have

$$N = 0{\times}2^7 + 1{\times}2^6 + 1{\times}2^5 + 1{\times}2^4 + 0{\times}2^3 + 1{\times}2^2 + 1{\times}2^1 + 0{\times}1 =$$
$$0 + 64 + 32 + 16 + 0 + 4 + 2 + 0 = 118$$

Table 10-1 lists the decimal equivalents of 2^n and 2^{-n}.

n	2^n	2^{-n}
0	1	1
1	2	.5
2	4	.25
3	8	.125
4	16	.0625
5	32	.03125
6	64	.015625
7	128	.0078125
8	256	.00390625
9	512	.001953125
10	1,025	.0009765625
11	2,048	.00048828125
12	4,096	.000244140625
13	8,192	.0001220703125
14	16,384	.00006103515625
15	32,768	.000030517578125
16	65,536	.0000152587890625

Table 10-1. Decimal Equivalents of 2^n and 2^{-n}.

Full Scale and Resolution

Figure 10-1 shows the input-output association for a 3-bit A/D converter with a full scale (F.S.) voltage of 10 volts. For a 3-bit code, a total of $2^3 = 8$ digital output codes are allowed. The resolution of this quantization is F.S./2^3 = 10/8 = 1.25. Thus, 1 LSB = 1.25 volts. In general, the resolution on an n bit converter using a natural binary coding is

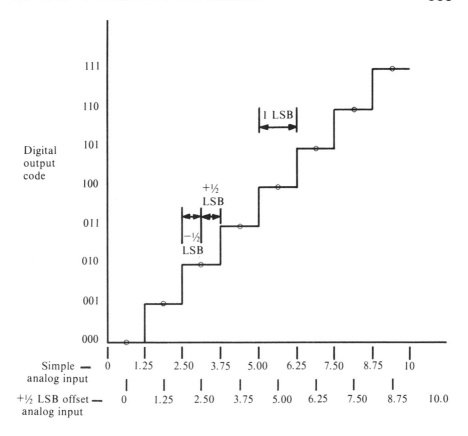

Figure 10-1. 3-bit A/D converter natural binary coding

$$1 \text{ LSB} = \frac{\text{F.S.}}{2^n} \qquad (10\text{-}2)$$

Note that no digital code exists for the full-scale voltage itself. This is a conventional practice and has the advantage that converters with identical Full-Scale voltages but varying bit number n will differ only in resolution; the cardinal voltages such as ¾ F.S., ½ F.S., ¼ F.S. remain unchanged (see Table 10-2). The maximum output voltage under this condition is F.S.–1LSB.

As shown in Figure 10-1, transitions from one digital number to the next occur at integral multiples of the LSB voltage. By offsetting the input voltage +½ LSB, the error band can be placed about the integral multiplier of the LSB. For example, if the digital output is 100 and the analog input has no offset, then we say the input voltage is 5.00 with a possible error of +1LSB, since the digital output is identical for the input range 5.00 to 6.24

		10V F.S.	
no of bits n	LSB	analog input for all 1's output (FS-1LSB) volts	½ scale
3	1.25v.	8.750±.625v.	5.0±.625v.
4	.625v.	9.375±.3125v.	5.0±.3125v.
6	.15625v.	9.8438±78.1mv.	5.00±78.1mv.
10	9.7656mv.	9.99023±4.88mv.	5.0000±.488mv.

Table 10-2. 10 Volt Full-Scale n-bit converter analog input offset + ½ LSB.

volts. However, if the analog input is offset +½ LSB, then 100 will correspond to 5.00 ±½ LSB. Of course, the *total* uncertainty is always 1 LSB.

10.2.1   Digital Codes

In certain A/D converter applications, it is desirable to represent the analog voltage with a digital code other than the natural binary representation treated above. The characteristics of some popular codes are discussed below.

Offset Binary — In the previously covered straight binary schemes, only unipolar input signals could be converted. In offset binary conversion, however, the input analog signal is offset by $- FS/2 + \frac{1}{2}(FS/2^n)$ so that the transition from the digital value 100...00 to 100...01 is made when the analog input voltage is $+ \frac{1}{2}(FS/2^n)$. Note that $\frac{1}{2}(FS/2^n) = \frac{1}{2}$ LSB for an n bit converter and that 100...00 will correspond to zero volts input.

Sign Magnitude — In sign magnitude conversion, the most significant bit represents the polarity of the analog voltage and the remainder of the bits represent the magnitude of the signal. In many systems such as panel meters, the fact that this scheme has two representations for zero i.e. 100...00 and 000...00 is not a limitation.

BCD Codes — Especially in digital display applications a BCD (see Chapter 9) coded format can result in considerable savings of decoding logic since display interfacing is straightforward. Unfortunately, only a few A/D conversion techniques are readily adapted to the BCD output scheme (see Section 10.4.). Of course, a natural binary code can be converted to BCD using a Binary to BCD parallel conversion device such as the 54/74185TTL I.C. It should be remembered that the BCD format is not an efficient method of storing digital information since this code requires more digital bits to represent numbers greater than 10 than does the straight binary coding scheme.

2's Complement — This notation is ideal for subsequent digital processing since arithmetic operations are easily performed (see Chapter 9). This format is simply obtained by inverting the most significant bit of the offset binary converter previously discussed. Again, the input is offset so that the transition 000...00 to 000...01 occurs at ½ LSB.

10.2.2 Sources of Error

The resolution error previously discussed, i.e. $FS/2^n$ is an essential quantization error and can only be reduced by increasing the number of bits n. In practical converter circuits additional errors can accumulate due to component matching problems and temperature effects. These errors can be classified as follows:

Linearity — This specification can be defined two ways: 1) as the maximum deviation from a straight line drawn between the F.S. output and zero, or 2) as the maximum deviation from the "best fit" straight line drawn

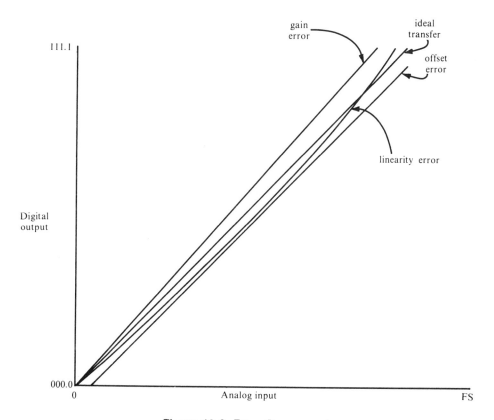

Figure 10-2. Error Components.

through the transfer function. Method 1 results in stricter control of nonlinearities. A good converter should have a maximum nonlinearity of $\pm$ ½ LSB. See Figure 10-2.

Gain Error; Offset Error — See Figure 10-2. Both of these effects are easily nulled with most converters. However, the temperature dependencies of these effects are not easily compensated.

Monotonicity — This specification requires that an increasing analog input result in an increasing digital output with no missing codes and vice versa for decreasing inputs. Manufacturers simply state that the converter is monotonic over its full range. In digital-to-analog converters lack of monotonicity implies that at some transition of the digital input to a higher value, the analog output will decrease instead of increase and vice versa.

10.3 DIGITAL TO ANALOG CONVERTERS

The D/A converter has many stand-alone uses such as a digitally controlled display driver or a servo positioning system, but its primary use is in conjunction with other circuit elements to provide high-accuracy A/D conversions. To a large degree, the performance of such A/D systems is determined by the performance of the D/A converter itself. Several popular D/A conversion schemes are treated below.

10.3.1 Weighted Current Sources

Figure 10-3(a) shows a simple 4-bit D/A converter. Since the inverting input of the operational amplifier is at virtual ground, the output voltage V_o is proportional to the current I created by switching one or more of the resistors R_1 - R_4 into the circuit. I is given by

$$I = a_1 \frac{V_s}{R_1} + a_2 \frac{V_s}{R_2} + a_3 \frac{V_s}{R_3} + a_4 \frac{V_s}{R_4} \qquad (10\text{-}3)$$

where $a_1 = 1$ if S_1 is closed and $a_1 = 0$ if S_1 is open and similarly for a_2, a_3, a_4 and switch S_2, S_3 and S_4 respectively. Therefore, the state of the switches S_1 – S_4 determines the analog output voltage. As shown, the resistor values provide a natural binary weighting of the switch states with S_1 representing the most significant bit. To create this weighting, each bit must provide a current to the summing line which is exactly one-half the current provided by the next more significant bit. This technique is useful for a small number of bits, but for larger bit numbers the weighting resistors have a very wide

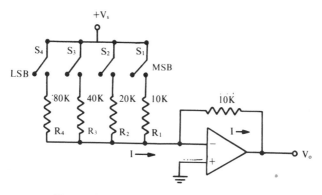

Figure 10-3. (a). Simple D/A converters.

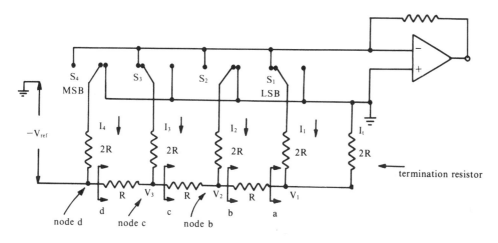

Figure 10-3b. D/A conversion using an R-2R ladder network.

range of values. For example, an 8-bit converter with a 10K MSB weighting resistor would have a 1.28M resistor in the LSB position. Unfortunately, it is very difficult to achieve temperature tracking for resistors of such different values. The R–2R ladder network overcomes these limitations.

10.3.2 R-2R Ladder Network

In Figure 10-3b, an R-2R ladder is shown in a current-mode connection. Switches S_1-S_4 connect the 2R resistors (except the termination resistor) either to ground or to the virtual ground of the summing amplifier and therefore, the d.c. currents in the ladder are unaffected by the switch states. To understand the ladder operation, consider a negative reference voltage $-V_{ref}$ attached to the ladder as shown. This will set up d.c. voltages V_3, V_2 and V_1 and d.c. currents I_4, I_3, I_2, I_1 will flow. Looking to the right of bracket (⤷)a we see the voltage V_1 across equal resistors setting up *equal* currents I_1 and I_t. The equivalent resistance to the right of bracket b is 2R and therefore with V_2 applied across equal resistors (2R in parallel with 2R), we may write

$$I_2 = I_1 + I_t \qquad \text{but } I_1 = I_t \qquad \text{so}$$
$$I_2 = 2I_1$$

Again, the total resistance to the right of bracket c is 2R and therefore I_3 equals the total current flowing into node c from the right, thus

$$I_3 = I_2 + 2I_1 \qquad \text{but } I_2 = 2I_1 \text{ so} \tag{10-4}$$
$$I_3 = 4I_1$$

Similarly, the resistance to the right of node d is also 2R and again I_4 equals the total current entering node d from the right so

$$I_4 = I_3 + 4I_1 \qquad \text{but } I_3 = 4I_1 \text{ (Equation 10-4) so}$$
$$I_4 = 8I_1$$

Clearly, a binary weighting scheme for I_4-I_1 results and the ladder could be extended any length to the left. The termination resistor is absolutely necessary since it "fools" the ladder into thinking that the R-2R resistor network extends infinitely to the right. Since only two resistor values are required, this scheme is easily expanded to large bit numbers with the assurance of tight temperature tracking. Furthermore, since only resistance ratios need be controlled the R-2R ladder is easily included into diffused integrated circuits.

Figure 10-4a illustrates the use of the ladder network in the current mode. This integrated circuit is the Motorola 1408L-8. A simplified current switch appears as Figure 10-4(b) and operation is as follows. A precision current source pulls a negative current I_{ref} out of the extreme left side of the ladder networks, thus setting the full scale output current. With the feedback connection shown between Q_2 and Q_3, the emitter of Q_2 will be exactly the base emitter drop of Q_3 above the reference voltage V_{ref}. Since the Q_3 of each stage is driven by an identical current source, I_c, and their emitters are all tied to V_{ref}, their base voltages will be identical. This guarantees that each leg of the R-2R network will be tied to an identical voltage. Furthermore, the base currents i_b of Q_2 and Q_3 tend to cancel each other ensuring that the precision current, I_n will flow unaltered in the collector of Q_2. Then if Q_1 is off, I_n will be steered to the output line I_o by diode D_1. If Q_1 is on, the collector of Q_2 will be pulled above the voltage of the output line and Q_1 will supply I_n to Q_2 thus removing I_n's contribution to the total output current I_o.

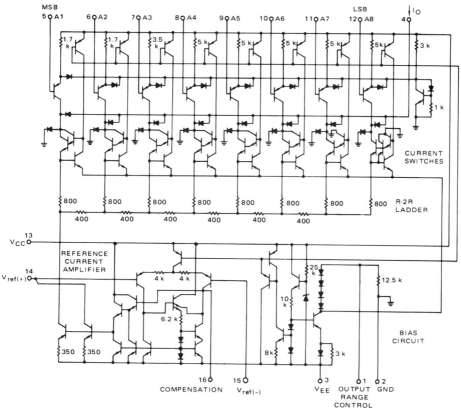

Figure 10-4a. "a" circuit diagram of the Motorola 1408L-8 Digital to Analog converter. (Courtesy of Motorola Semiconductor Products, Inc.)

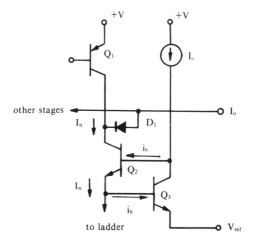

Figure 10-4b. Simplified current switch.

One other interesting circuit function is that performed by the extra transistor in the LSB current switch. As mentioned, an 800Ω termination is required for R-2R ladder operation, but this implies that equal currents will flow in the last two ladder resistors on the right. In this design, the LSB current is divided in half by splitting the current with parallel connected transistors and shunting one collector to ground as shown. This extra current is often called the remainder term and its loss causes the full-scale output current to be equal to I_{ref}-1LSB.

10.3.3   Settling Time

When the digital input code changes, some period of time, t_s, will be required for the analog output parameter (voltage or current) to adjust itself to the correct value. t_s is called the settling time and defined as the time required for the analog parameter to settle within $\pm\frac{1}{2}$ LSB following the input code change. The most stringent requirements will be placed on the MSB since $\frac{1}{2}$ LSB represents a small percentage of its total change. For example, when switching the MSB on in a 2,000 ma full-scale 8-bit D/A, the output must settle to within 4μa of 1.00mA to achieve 8-bit accuracy. However, the LSB of the same converter (weight 7.8μA) need only reach 4μA when switched on to achieve similar accuracy. In the MC1408 just discussed, MSB settling occurs in about 300 ns for the on transition, but due to nonsymmetric switch design the MSB can be switched off in about 80 ns.

Any circuitry used to sense or amplify the D/A output can only degrade the settling time. As a first approximation, this circuitry can be modeled as a single pole system. Figure 10-5(a) shows the number of time constants required for a single pole system to settle to within a given percentage of the final value in response to a step input. As an example, consider the circuit of Figure 10-5(b) which shows a resistively loaded, current output D/A converter. C_s represents the total stray capacitance and is composed of capacitance within the converter itself and any additional circuit capacitance contributed by board layout and/or a sensing device such as an oscilloscope probe. For an 8-bit converter, the output must settle to within 0.2% (=$\frac{1}{2}$LSB) of its final value following a transition of the MSB. If the basic converter settles within 80 ns, then it is reasonable to require that the external circuitry add only an additional 20 ns to t_s. From Figure 10-5(a) we see that 0.2% settling for the single pole system will require about 6 time constants and therefore we require that

$$RC \leq \frac{20 \text{ ns}}{6} = 3.3 \text{ ns} \qquad (10\text{-}5)$$

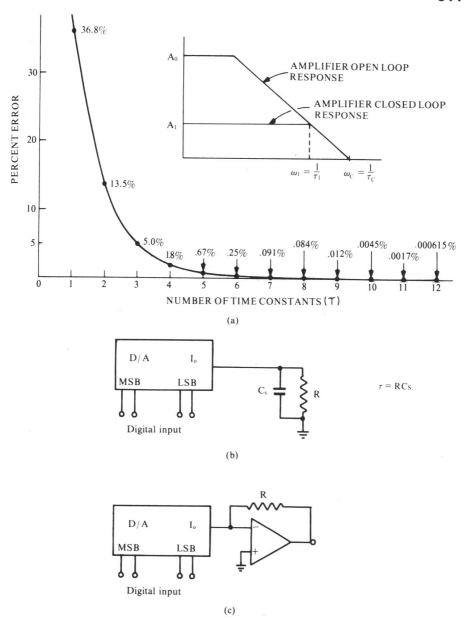

Figure 10-5. Interfacing the output of a Digital to Analog converter.

If C_s is 15 pf, a typical value for a high-speed converter, then Equation (10-5) sets the maximum allowable value of R at 220. Unfortunately, this low value of resistance will result in an unacceptably low output voltage for many applications. Fortunately, a high-speed circuit with sufficient voltage gain can be achieved by feeding the current output directly into the summing junction of a high-speed operational amplifier connected in the inverting mode (see Chapter 5). This configuration, Figure 10-5(c), provides both low input resistance and voltage gain.

From the preceding discussion, we can appreciate the difficulty of designing an accurate high-speed D/A converter. Fortunately, considerable effort has been expended by the I.C. manufacturers in recent years to design and market low-cost, high-performance integrated D/A circuits, thereby freeing the system designer from the task of building a discreet D/A converter. However, a thorough understanding of D/A principles will always be a prerequisite to effective use of these building blocks.

10.3.4   Some Notes for Monolithic D/A Converter Users

Although not technically considered performance specifications, the following characteristics must be considered when specifying integrated D/A converters.

Output voltage compliance—For those converters with current outputs, there is a limited range of voltages over which the output line can operate and still maintain the specified accuracy. This voltage range is called the output compliance and is usually expressed in volts above and below circuit ground potential. Some converters have extremely limited compliances; for example the MC1408 has a +0.50 and –0.6 volt voltage compliance, whereas others, such as the Precision Monolythics, Inc. DAC-08, have compliances as high as +18 volts and –10 volts. Furthermore, D/A's that can operate over a wide supply voltage range usually have compliances which are a function of the supply voltage.

Usually, a converter is designed to have inputs which are compatible with one logic family such as TTL or CMOS. However, some converters (e.g. DAC-08) have a separate control line which allows the input threshold voltage to be varied allowing interface with any logic family including ECL.

Glitches and deglitching — Glitches in the analog output can occur at any transition of the digital input, but they are most annoying at the cardinal transitions. For example, consider the transition to 1/2 scale for the 6-bit D/A depicted in Figure 10-6. With the input code at 011111 all lower current switches are turned on. When the code advances to the next binary number, i.e. 100000, the lower-order switches may not turn off as quickly as the MSB can turn on. Thus, for a moment, the input code appears as 111111 causing the inverting buffer amp to go quickly positive until the lower-order switches fully turn off resulting in the "glitch" shown in Figure 10-6(a). Note

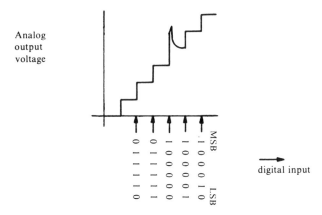

Figure 10-6a. Glitch at cardinal transistion for a 6 bit 2 volt F.S. D/A converter measured at point a.

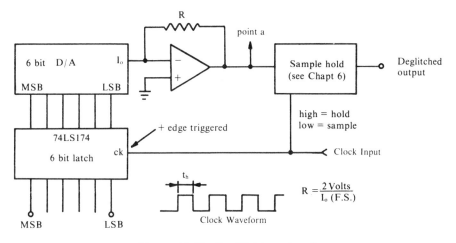

Figure 10-6b. Deglitching circuit.

that the width of the glitch is approximately the difference between the turn-on time and the turn-off time of the current switches themselves and is not related to settling time. Glitches are most annoying in D/A converter CRT display applications where the glitch can be seen for what it is and in precision servo control applications where a sensitive control loop can be upset by the voltage spike. A sample-hold circuit can be used to deglitch a D/A in critical applications by holding the last stable value during the digital input transition and acquiring the new value at least one settling time t_s later. In the circuit of Figure 10-6(b), performance can be optimised by setting the clock high time t_h to be slightly greater than the combined settling times of the D/A and inverting amplifier. Deglitching is achieved at a price of adding a fixed delay equal to t_h, to the overall transfer function of the converter. Care must be taken in the overall system design to insure that the digital inputs to the 74LS174 are stable during the positive clock transition.

10.4 ANALOG TO DIGITAL CONVERSION

In contrast to D/A conversion, a variety of A/D conversion techniques are available. However, four types have established themselves as industry standards, they are:
1) Integrating types.
2) Tracking converters.
3) Successive approximation types.
4) Parallel type.

The above approaches represent varying degrees of trade-off between resolution, conversion speed and cost. As noted, some types will require a sample-hold circuit at the converter input.

10.4.1 Integrating types

The integrating type offers the highest resolution and the lowest cost. In addition a sample-hold circuit is not usually required. The principal disadvantage is the relatively long conversion time, typically several milliseconds.

Figure 10-7 shows a simple single-ramp system. Assuming that the flip-flop and the counter are cleared, operation begins when a start pulse occurs. At this time the integrator begins a positive-going ramp and the counter is enabled. When the ramp reaches the analog input voltage the comparator clocks the flip-flop, resetting the ramp and stopping the counter. Since the length of time required for the ramp to reach the analog input is proportional to the input voltage, the digital count left on the counter is also proportional to the analog input. For a clock frequency, f, we have

$$\text{Digital count} = \frac{V_{in}CR}{V_{ref}} f(Hz)$$

Unfortunately, the count is proportional to R, C, V_{ref} and f, and is therefore sensitive to variations in any of these parameters. The dual-slope converter overcomes these limitations.

Dual slope integrating A/D converter—The dual-slope converter, Figure 10-8, requires more control logic than the single-slope converter, but has the advantage that the ultimate accuracy is determined only by the

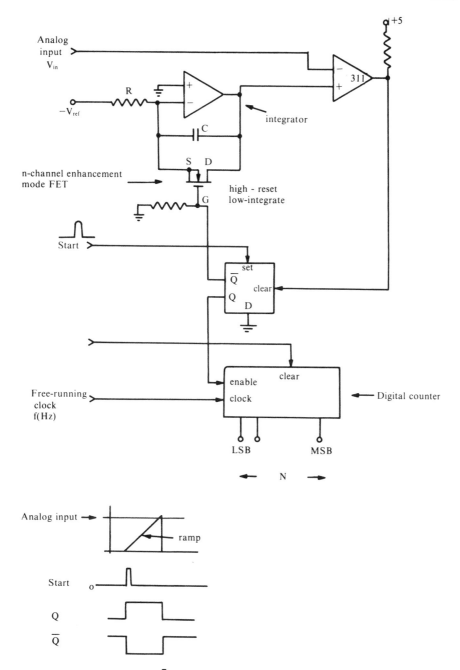

Figure 10-7. Simple integrating type Analog to Digital converter.

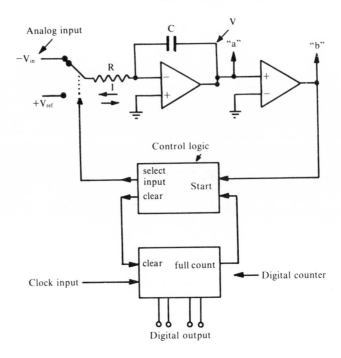

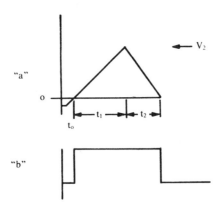

Figure 10-8. Simplified dual ramp converter.

stability of the reference voltage. The waveforms shown in Figure 10-8 result from the following sequence. The digital counter is initially held clear and the unknown analog input V_{in} is applied to the integrator which starts the output slewing positive at a rate equal to

$$\frac{\Delta V}{\Delta t} = \frac{I}{C} = \frac{V_{in}}{RC} \qquad \textbf{(10-6)}$$

From a previous conversion, point a was left below ground, but at time t_o the integrator output crosses the zero level beginning the time period t_1. At t_o, the digital counter begins counting and continues until the counter is full, N_{max}, marking the end of period t_1. At this point the control logic clears the counter and connects the integrator input to the positive reference V_{ref}, causing the output to begin slewing negative at a rate equal to

$$\frac{\Delta V}{\Delta t} = \frac{V_{ref}}{RC} \qquad \textbf{(10-7)}$$

The counter is enabled throughout the period t_2 and when the integrator reads zero, the digital count N is held.

The peak voltage on the integrator V_2 is found from Equation 10-6 thus

$$V_2 = t_1 \frac{V_{in}}{RC} \qquad \textbf{(10-8)}$$

and t_2 is found from Equation 10-7 as

$$t_2 = \frac{V_2 RC}{V_{ref}} \qquad \textbf{(10-9)}$$

Substituting for V_2 in Equation 10-9 we have:

$$\frac{t_2}{t_1} = \frac{V_{in}}{V_{ref}} \qquad \textbf{(10-10)}$$

Furthermore, we can write

$$N_{max} = t_1 f_{clock} \quad \text{and } N = t_2 f_{clock}$$

and from Equation 10-10 we have

$$\frac{N}{N_{max}} = \frac{t_2}{t_1} \times \frac{f_{clock}}{f_{clock}} = \frac{V_{in}}{V_{ref}} \qquad \textbf{(10-11)}$$

and therefore

$$V_{in} = \frac{N}{N_{max}} V_{ref} \qquad \textbf{(10-12)}$$

We see that if R, C and f_{clock} are stable over the conversion period t_1 + t_2, then the cancellations indicated in Equation 10-10 and Equation 10-11 are valid. Note that any comparator input offset is also canceled since offset has an equal effect on t_1 and t_2. A remaining source of error, however, is the integrator input offset voltage, V_{os} since this affects the charging current I in a way that does not cancel over the conversion period. At the expense of some additional control logic, this error can be eliminated by the circuit of Figure 10-9. Prior to the beginning of a conversion, Q_1 and Q_2 are held "on" which connects the operational amplifier in the voltage-follower mode and grounds the right side of the input resistor R. With a low impedance path on both terminals of C_1, it quickly charges to a voltage equal to V_{os}, but opposite in polarity. When Q_1 and Q_2 are turned off and the conversion begun, C_1 cancels the effect of V_{os}. Naturally, C_1 must be sufficiently large so that the input bias current of the operational amplifier will not appreciably alter the charge on C_1 during the conversion period.

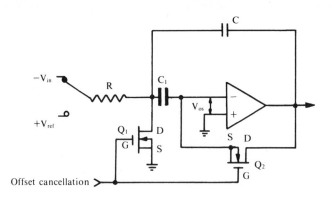

Figure 10-9. Offset cancelling circuit for the integrator in Figure 10-8.

A further advantage inherent in ramp-type convertors is the capacity to generate any digital counting sequence (binary, BCD, Gray, etc.) simply by choosing the appropriate counter. It is particularly advantageous to use BCD registers when the A/D output is used to drive a digital numeric display.

10.4.2 Tracking Converters

A simple tracking A/D is shown in Figure 10-10(a). The binary counter is clocked continuously and the count either increases or decreases, depending on the state of the count-up/count-down control. The response of this system to a step input is illustrated in the timing diagram in Figure 10-10(b). As long as I_{in} is larger than I_o, the counter increases one bit at each positive clock transition causing the D/A output current to slew negative as shown. At 01001 I_o is larger than I_{in} and so the comparator output goes low. On the next positive clock transition the counter increments back to 01000 setting I_o slightly less than I_{in} so the comparator goes high again and so on.

We see that as long as the slew rate limit is not exceeded, this system tracks, within $\pm$ 1LSB of the desired value.

For a clock frequency f we see that the current slew rate limit is given by

$$\text{Slew rate} \quad \frac{(\text{amp})}{(\text{sec})} = 1 \text{ LSB} \times \text{f} \qquad \textbf{(10-13)}$$

Assuming identical full-scale currents and a fixed clock frequency, f, Equation 10-13 states that increased slew rate can only be obtained by decreasing the resolution. Similarly, higher resolution demands reduced slew capability. The clock frequency should be set as high as possible since this allows increased slew rate with no loss of resolution.

For an n bit converter, $2^n/\text{f}$ seconds are required for the converter to

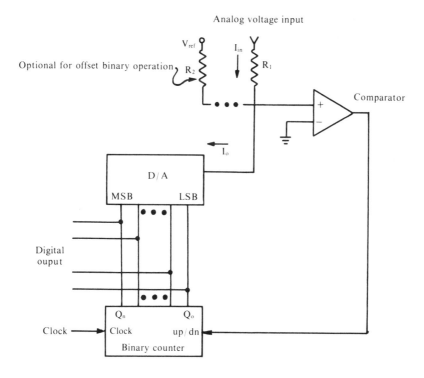

Figure 10-10a. Tracking A/D converter.

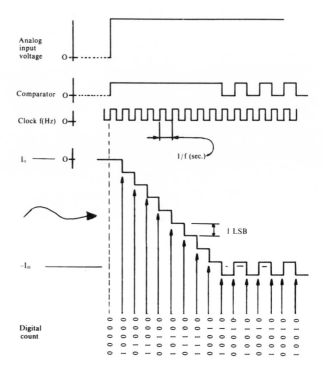

Figure 10-10b. Tracking A/D wave forms.

acquire an analog input transition equal to full scale. An 8-bit tracking converter operating at a 2MHz clock rate would therefore require $256/2\text{MHz} = 128$ μsec to "lock in" to a worst case (full scale) input transition. This is a severe speed limitation in multiplexed or high-speed sampling systems where the unique tracking feature of this converter is not of particular advantage. The tracking converter does not require a sample-hold circuit at its input.

10.4.3 Successive Approximation

The successive approximation (SA) A/D converter offers a very significant speed advantage over the tracking type with only a slight increase in system complexity and cost. As a result, the SA converter has established itself as the most popular converter type in moderate to high-speed (100 μs to 1 μs) systems. In contrast to the tracking type, SA converters must always be preceded by a sample-hold circuit to insure that the input does not change during the conversion period.

A simple guessing game provides a good analogy to the operation of the SA converter. Suppose I have a number in mind between 0 and 64 and you try to guess this number by asking questions to which I can answer only "yes" or "no." One could simply start at 0 and ask is the unknown greater than 0; is it greater than 1; is it greater than 2 and so until the unknown value was reached. This is exactly the approach taken by the tracking A/D converter and, as we saw, for a "worst case" unknown (i.e., 64) we would have to ask 64 questions. The SA converter works in a more sophisticated fashion. For example, let the unknown number equal 54. Our first question will be "is the unknown greater than 32?" The answer is "yes" so we divide the remaining band (32-64) in half and ask "is the unknown greater than 48?" Again the answer is "yes" so again we divide the remaining band (49-64) in half and ask "is the unknown greater than 56?" This time the answer is "no." Proceeding in this fashion we would arrive at the unknown integer having asked the fewest possible questions. Specifically, if the largest allowable number can be represented as 2^n; then n questions are required to reach the unknown integer. In our example, the largest possible number is 64 and since $2^6 = 64$ we need ask only 6 questions to arrive at the unknown value.

Figure 10-11 shows a 6-bit SA converter and typical waveforms. By comparison with our guessing game we see that the control logic generates the questions in digital form and the D/A converts this number into a current which is compared with the unknown current, V_{in}/R. The comparator answers the question, "Is this number greater than the unknown?" and supplies this information to the control logic so that

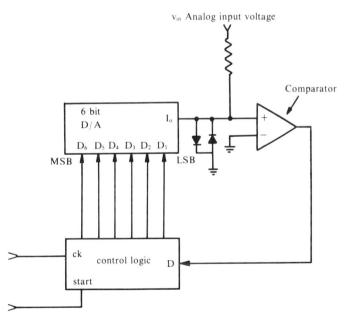

Figure 10-11a. Block diagram 6 bit Successive Approximation A/D connector.

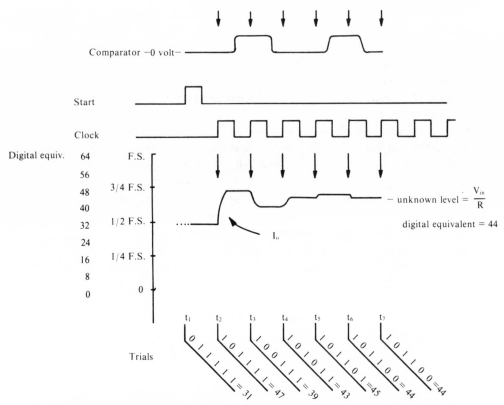

Figure 10-11b. 6-bit Successive Approximation converter with unknown
equal to $^{44}/_{64} \times$ F.S.

subsequent trial questions can be determined. To generate the correct set of
questions, the control logic for an n-bit converter must perform the
following operations based on the comparator's decision:

Trial No. 1 - a) Set bit n - low
             b) Set remaining bits - high

Trial No. 2 - a) Bit n left low if trial No. 1 $>$ unknown
             a′) Bit n set high if trial No. 1 $<$ unknown
             b) Set bit n-1 low

Trial No. 3 - a) Bit n-1 left low if trial No. 2 $>$ unknown
             a′) Bit n-1 set high if trial No. 2 $<$ unknown
             b) Set bit n-2 low

In general (except for trial No. 1)

Trial m - a) Bit n-(m-2) left low if trial (m-1) $>$ unknown
          a′) Bit n-(m-2) set high if trial (m-1) $<$ unknown
          b) Set bit n-(m-2) low.

In the example of Figure 10-11b, an input current, V_{in}/R equal to $44/64 \times$ F.S. is applied to the circuit of Figure 10-11a and the conversion sequence is shown in detail. At time t_1 the D/A input is set at 1/2 scale-1 LSB i.e., $31/64 \times$ F.S. and the comparator indicates that this value is less than the unknown current. Therefore, for trial 2, time t_2, the MSB (bit n) is set high and the next most significant bit (bit n-1) is set low. The remainder of the conversion proceeds according to the rules outlined above with each subsequent trial being performed on the positive transition of the clock. As would be the case in a practical high-speed SA converter, the response delays of the D/A converter and comparator are clearly visible in Figure 10-11b.

As mentioned, the number of trials required is n and so with a clock frequency of f(Hz) the conversion rate is; f/n conversions/second. Four factors determine the maximum clock frequency, thereby setting an upper limit on the conversion speed of an n bit system:

1) Delay time — clock transition to control logic outputs valid
2) D/A settling time
3) Delay time — comparator input to output
4) Setup time — control logic. This is the length of time that the comparator data must be valid at the control logic input prior to the positive clock transition.

The control logic section is available as a TTL integrated circuit from several manufacturers; for example the Advanced Micro devices series AM 2502/3/4 or National Semiconductor series DM 2502/3/4. With these circuits an 8-bit 3 I.C. Successive Approximation converter operating with a clock frequency of about 2MHZ can be constructed. If the 2502/3/4 series is replaced with a control logic section which uses Schottky TTL or ECL logic gates, clock frequencies in excess of 5 MHZ can be achieved using a high-speed D/A such as the DAC-08. In the present marketplace, the availability of higher-performance SA converters is limited only by demand, since 8, 10 and 12-bit SA converters which operate at 50 MHZ clock rates have been supplied to the space program by TRW Inc. In addition, these converters contain the entire circuitry of Figure 10-11a integrated on a single chip.

10.4.4   Parallel Conversion Technique

By far the fastest conversion technique, the Parallel converter can perform an entire conversion at a 100 MHZ rate (see Figure 10-12), Circuit operation is quite simple. The resistor string sets the inverting input of each comparator to a slightly higher voltage than the previous converter. Thus, for any input voltage, V_{in}, within the full-scale range, comparators biased

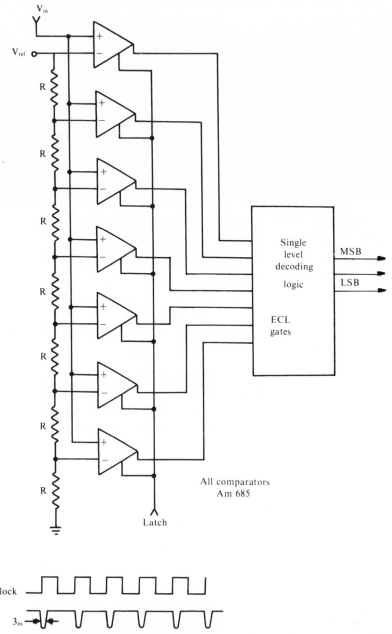

Figure 10-12. Parallel A/D Converter.

below V_{in} will have high outputs and comparators biased above V_{in} will have low outputs. Unfortunately, the comparator outputs are not coded in natural binary and a decoder is necessary if a binary output is desired.

Several high-speed comparators with propagation delays less than 6 nanoseconds are available (e.g., the Advanced Micro Devices Am 685), and therefore extremely fast conversion speeds are possible.

Since most very high-speed comparators have a latch input which holds and transfers the input state to the outputs, a sample hold circuit is not required at the input to the parallel converter if the comparators are latched simultaneously as shown in Figure 10-12.

Unfortunately, the number of comparators required for an n bit conversion is $2^n - 1$. In Figure 10-12, the 3-bit converter shown requires $2^3 - 1$ = 7 comparators. A 4-bit converter would require $2^4 - 1 = 15$ comparators and so on. By weighting the biasing resistors logarithmically, a "percentage accurate" coding will result, somewhat offsetting the coarseness of the small bit number conversion.

A hybrid combination of parallel and successive approximation techniques will yield an intermediate speed system. Figure 10-13 shows a block diagram of such a system. The first 3-bit converter yields the first 3 most significant bits of the final 6-bit word. These bits are applied to a D/A

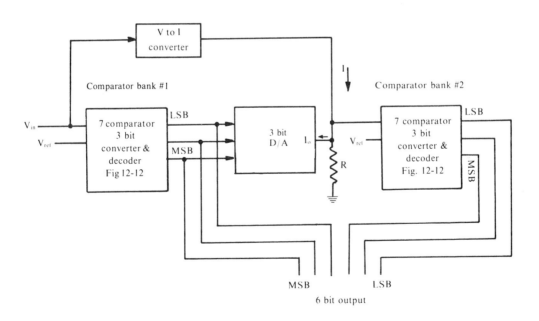

Figure 10-13. 10 MHz 6 bit D/A converter using a hybrid parallel/successive approximation technique.

and this result is subtracted from the original signal, the difference, scaled by resistor R is applied to the input of a second parallel 3-bit converter which generates the remaining less significant bits. The basic speed limitation in this system is the 3-bit D/A. However, for a high-speed 8-bit D/A such as the DAC-08 we could expect the first most significant bits to settle to 3-bit accuracy within about 60 ns. Since the remaining elements of this circuit are quite fast, 6-bit conversions at a 10MHz rate could be expected from this system. With latches on the first 3 bits and some analog switching, one bank of comparators could be eliminated. Unfortunately, any of the hybrid techniques require a sample-hold circuit at the input to maintain the input constant during the conversion period.

Chapter 11

Grounding and Shielding

11.1 When a circuit design is first realized in hardware form (a breadboard or a prototype) it becomes subject to several types of problems which are difficult or impossible to anticipate by analyzing the circuit diagrams. Broadly categorized, these problems are caused by 1) the effects of external electric, magnetic or electro-magnetic fields in sensitive portions of the system, and/or 2) the coupling of unwanted signals from one section of the system to another through either an ohmic connection such as a ground lead or via stray fields as mentioned above. In an analog system, these effects are observed as spurious, perhaps intermittent, signals mixed in with the predicted or desired signals and in a digital system the symptoms are usually false triggering of clocked devices (e.g., flip-flops and counters) resulting in degraded or useless output data.

The development of increasingly complex I.C.'s has made some aspects of the system designer's job more straightforward, but this same development has greatly complicated the task of eliminating unwanted signals from sensitive sections of a complex system. The analog system designer is specifying lower signal levels, wider bandwidths, higher impedances and sensitivities while the digital designer is demanding ever-increasing system speeds. These trends combined with higher circuit densities and increasingly polluted (electrically speaking) operating environments have elevated the task of controlling these unwanted signals to an unprecedented level of difficulty. As if the problem wasn't bad enough, the growing popularity of the hybrid analog-digital system has added a few new twists of its own since, as we shall see, the rules for the proper layout of an analog system *are in conflict with* the proper layout rules for a digital system!

In this chapter, we shall develop solutions to our stray signal problems through an understanding of the physical mechanisms which are typically responsible for coupling those signals into our system. With this knowledge, the designer will be able to anticipate many problem areas before his design reaches the breadboard stage and he will have the tools to effect a cure for those troubles which become apparent only at the prototype level.

We have avoided applying the term *noise* to the unwanted signals of this chapter preferring to reserve this nomenclature for signals of the type discussed in Chapter 3. There must, of course, be some distinction since the *noise* treated in that chapter is not a spurious effect at all, but rather it is a fundamental and well-defined characteristic of active and passive devices. Using the results of Chapter 3, the designer may determine the effects of electronic noise on his final design much as he would account for the effect of input offset voltage or current. There is little excuse for being surprised by the effect of electronic noise when the first breadboard is constructed. Therefore, in dealing with the somewhat more elusive problems that arise when unwanted signals are coupled into our system, we will use the terms *spurious* or *unwanted* in referring to the signals of the present chapter.

11.2   GUIDE LINES FOR ANALOG SYSTEMS

11.2.1   Ground Loops

The term *ground loop* is synonymous in the mind of many designers with spurious signal pickup. Certainly, the first words one hears upon the discovery of a contaminating signal is "you've got a ground loop." In fact, ground loops account for many instances of unwanted signal pickup, especially in analog systems, but unfortunately, they frequently become the scapegoat for all spurious signal problems.

The coupling mechanism associated with a ground loop is easily understood. Figure 11-1a shows a system comprising a microphone, a preamplifier, and a power stage driving a speaker. In a typical application of such a system the microphone and loudspeaker might be located close together and the preamp, poweramp, and power supplies would be remotely located. Wishing to save a length of wire, the novice engineer might interconnect this system as shown in Figure 11-1b using a common return line, l_1, for the microphone and the loudspeaker. Negligible current flows in the microphone, but the speaker current, i_{out}, may be several amps rms. This current flows in l_1 and because of the finite resistance of this line, $\dot{R}$, a voltage equal to $i_{out} \times R l_1$ is generated between node 1 (ground) and node 2.

Schematically, we must represent this effect by showing a voltage source equal to R × i_{out} in series with the microphone, (see Figure 11-1c). Clearly, this signal source represents an unwanted contamination of the microphone output. In this simple system, voltage v_{12} is not really spurious since it is related to the signal originating in the microphone and, therefore, we could categorize this as a type of unwanted feedback. In the worst case, there could be a frequency at which the overall unwanted feedback is positive and if sufficient loop gain existed, the circuit of Figure 11-1c would begin spontaneous oscillations. At best, feedback signal v_{12} will modify the overall gain and this effect will most likely be frequency-sensitive resulting in poor fidelity. We may easily imagine another speaker or power device connected to node 2 thereby using l_1 for a ground return line. If these additional signals are unrelated to our microphone signal then the effects of these signals may truly be called spurious. (In the preceding example we have ignored the possible *audio* feedback path which might exist between the speaker and the microphone. Even with well-conceived grounds, audio feedback could itself cause oscillations or uneven gain characteristics.)

The failure to account for the effect of finite lead resistance led to the ill-fated connection of Figure 11-1b. Any inductance associated with l_1 could only cause further problems. This effect is traditionally called a *ground loop* and it can always be physically explained by a diagram similar to Figure 11-1b and schematically modeled as shown in Figure 11-1c. Unfortunately, the term *ground loop* is not very descriptive since an obvious *loop* is not apparent in either figure. In the same vein, we may even question the use of the word "ground" in *ground loop* since, at the heart of the problem is the fact that the "ground" line is not equipotential. The term, however, has achieved considerable conventional status and we shall use it to describe the type of unwanted signal coupling most clearly described by Figures 11-1a and 11-1b.

11.2.2 Solving Ground Loop Problems

Separate Ground Return Lines

The most effective means of avoiding ground loops is to provide a separate ground return to the power supply for each grounded point in the system. Thus, by simply routing the microphone ground lead, l_2, as shown in Figure 11-1d we avoid the effect of the unwanted signal v_{12} caused by the speaker current flowing in wire l_1. If there are many grounded points within a system, then the number of separate ground leads required becomes unwieldy and we must find a suitable compromise. In practice, we need only supply a separate return lead for ground connections which are particularly

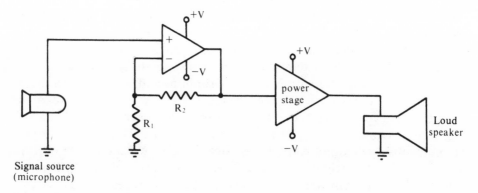

Figure 11-1a. Schematic representation of audio system.

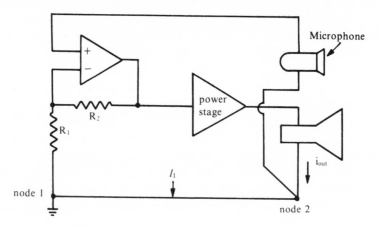

Figure 11-1b. Realization of Figure 11-1a showing common ground return line I_1.

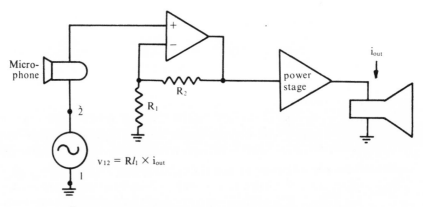

Figure 11-1c. Equivalent circuit of Figure 11-1b.

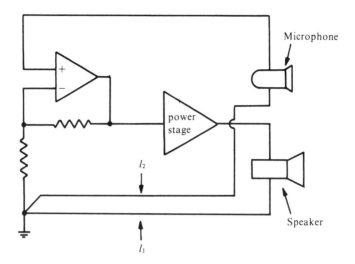

Figure 11-1d.

Figure 11-1. Figure 11-1a shows the circuit we wish to construct. Unfortunately, as shown in Figure 11-1b, the microphone ground lead is physically connected to the speaker ground lead coupling the unwanted signal generated between nodes 1 and 2 into the preamp as shown schematically in Figure 11-1c. The problem is alleviated in Figure 11-1d by providing a separate ground return I_2, for the microphone.

long or carry relatively significant current. It should be noted that providing separate ground returns is the fundamental solution to ground loop problems. Most other solutions provide somewhat degraded performance, especially at high frequencies as discussed below.

Utilize the Ability of an Operational Amplifier to Reject Common Mode Voltages

In many complex or multiple system applications it is either impractical or impossible to follow good ground practices througout the network. Consider the two-system network shown in Figure 11-2. Sound ground practices have been applied throughout systems 1 and 2 relative to nodes 1 and 2 respectively. However, due to some other constraint (perhaps they are located some distance from one another) separate ground lines, I_1 and I_2, are used to connect each system to the common power supply ground, node 3. Associated with nodes 1 and 2 are voltage v_1 and v_2 (measured relative to node 3) created when the ground return currents of each system, i_1 and i_2, flow through lines I_1 and I_2 due to the resistance and inductance of those lines. Currents i_1 and i_2 are composed of a.c. and d.c. components and since the internal operations of systems 1 and 2 may be

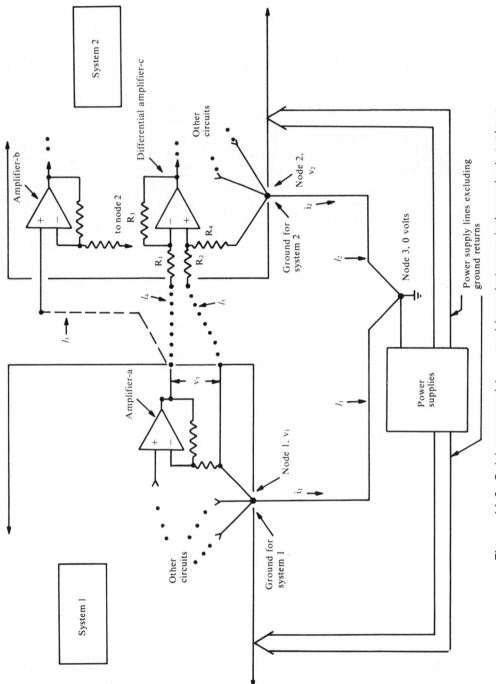

Figure 11-2. Solving ground loop problems between independent systems.

quite different, these two currents are largely independent. The voltage difference between nodes 1 and 2 is $v_2 - v_1$ and is also composed of a.c. and d.c. signals. Now, voltage v_1 and v_2 affect neither of the circuits in system #1 or #2 respectively since proper grounding procedures have been followed in each system. But, if one system is to communicate with the other system, then the voltage *difference* $v_2 - v_1$ will have serious consequences. Consider the output of amplifier a, Figure 11-2, connected to the input of amplifier b by the single dashed line l_3. (The dotted lines l_4 and l_5 are not connected.) If the input to amplifier a is zero volts relative to system #1 ground (node 1), then v_3 is zero volts and the voltage on line, l_3, is v_1. But amplifier b is referenced to node #2 (voltage v_2) and so the apparent input signal to amplifier b is $v_2 - v_1$. the voltage difference between nodes #1 and #2. Under these conditions, the output of amplifier b is the gain of this stage, A_b multiplied by $v_2 - v_1$, but, had the communication channel functioned correctly, the output should have been $v_3 \times A_b$, that is, zero volts. Clearly, this interconnection technique is unsatisfactory.

A viable solution to this problem utilizes differential amplifier c (see Chapter 1). Allow our system interconnection to be composed only of the dotted lines l_4 and l_5. We see immediately that v_3 becomes the *differential* input to amplifier c and $v_2 - v_1$ becomes a common-mode signal. Therefore, to the extent that amplifier c can reject the unwanted common-mode signal, $(v_2 - v_1)$, the integrity of our communications channel is restored. Under no circumstances should l_5 be connected to node #2 in system #2 since a portion of the ground currents i_1 and i_2 would flow through l_5 creating an unwanted voltage differential along its length. If the interconnection must travel long distances or through an especially hostile electrical environment, l_4 and l_5 should be the inner conductors of a shielded twisted pair cable such as Belden #9452 minature audio cable. Some slight performance differences may be observed depending upon whether the cable shield is connected to node #1 or #2, but it should be connected to one *or* the other. In some instances the performance of a particular channel may be improved by connecting the cable *shield* to both nodes 1 and 2, but this may affect the performance of other channels. (*Note*: this is quite different from connecting l_5 to both nodes.) Although degrading the performance of some communication links, coaxial cable may be used in place of the shielded twisted pair. In this instance l_4 must be the center conductor and l_5 the shield connected only to node #1 as shown in Figure 11-2.

Beware at Higher Frequencies

The effectiveness of the solution discussed in the previous section is directly related to the ability of the differential amplifier to reject the unwanted common-mode signals. The common-mode rejection ratio

CMRR of any differential amplifier is defined so as to allow us to calculate the equivalent *differential* input signal caused by a common-mode voltage on the input terminals. Therefore, we may evaluate the relative performance of several differential amplifiers *independent* of their respective differential voltage gains (see Chapter 1, Section 3.4). If for example, the CMRR of a particular differential amplifier is 80 db (a factor of 10,000), then a 1-volt common-mode signal will cause an equivalent differential *input* signal of $1/10,000 = 100 \mu$ volts. The effect of this common-mode signal on the *output* of this amplifier is found by multiplying $100\mu v$ by the differential gain of the amplifier.

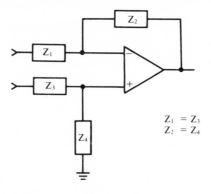

Figure 11-3a. To achieve op amp limited CMRR, impedances Z_1-Z_4 must be balanced as shown.

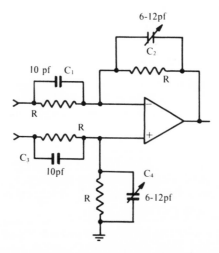

Figure 11-3b. A practical method of balancing impedances Z_1-Z_4.

Two factors determine the CMRR of the differential amplifier pictured in Figure 11-3a. They are:

1) The degree of matching of the impedance ratios Z_1/Z_2 and Z_3/Z_4. Thus, to achieve a given CMRR, say 60 decibles, the required degree of matching expressed as a percentage is $1/1,000\,(60db) \times 100 = .1\%$ for all signal frequencies. For d-c. signals we need to balance the resistance in the network to the desired percentage, but at higher frequencies the effects of stray capacitances will unbalance the impedances especially if larger resistor values are used. Since only a few one-hundredths of a percent impedance variation may degrade CMRR, we find that the effects of stray capacitance are evident at frequencies orders of magnitude below the 3 db down frequency associated with the network resistances and the stray capacitance.

Figure 11-3b shows a unity-gain differential amplifier. Since all resistors are of equal value, we may obtain a matched set and be assured of tight temperature coefficient matching as well. To swamp the effects of stray capacitance we have added NPO type ceramic capacitors around each resistor. Capacitors C_2 and C_4 must be adjusted to provide a frequency independent gain characteristic *and* optimize the CMRR.

2) If the network impedances are balanced as described above, the CMRR will be determined by the characteristics of the operational amplifier. As evidenced by Figure 11-4 the CMRR is maximum at d.c. and low frequencies and decreases at a 20dB/decade rate at higher frequencies. This curve represents the best performance that can be achieved using this amplifier and is obtainable only with exceptional matching of the external network impedances and low source resistances.

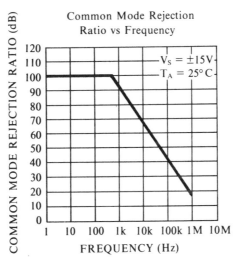

Figure 11-4. CMRR of a typical operational amplifier.

11.2.3   Rejecting Signals on the Power Supply Lines

As described in Section 1.3.5, the power supply rejection ratio PSRR is a measure of an operational amplifier's ability to reject signals which are present on the power supply lines. In a given application, we may easily calculate the maximum allowable power supply signals from the information provided in Chapter 1 and the PSRR versus frequency curve for the particular amplifier. There is, however, a less obvious consequence of finite PSRR.

The very existence of a PSRR implies that the voltage supply terminals of any operational amplifier may be treated as an input. Although the phase relationship between inputs and outputs for this type of signal is unspecified we may assume that at least one supply terminal may be treated as a noninverting input. Consider the circuit of Figure 11-5. The current which the unity voltage gain power stage supplies to the load R_L must flow through the power supply wiring resistance Rl_1 or Rl_2 creating a.c. signal v_1 and v_2. At some frequency, v_1 or v_2 will have the proper phasing to provide overall positive feedback via one of the operational amplifier power supplies and if the gain of the operational amplifier is sufficiently high, oscillations will result. Notice that we have included the 0.1 μF capacitors usually recommended by device manufacturers at the operational amplifier supply terminals. However, because of the low line resistance Rl_1 and Rl_2 and the high load currents, these capacitors have little decoupling effect. Two straightforward solutions to this type of problem are shown in Figure 11-6. Since the PSRR ratio of a typical operational amplifier is a decreasing function of frequencies the problems described above are most evident at

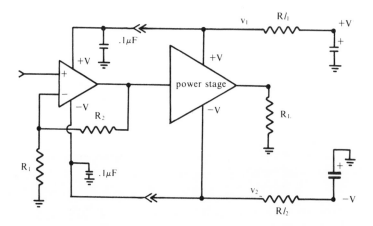

Figure 11-5. A power supply interconnection technique which may result in circuit oscillations.

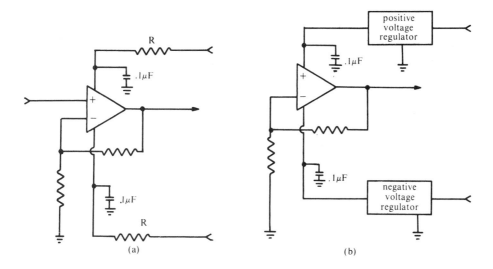

Figure 11-6. Controlling possible positive feed back paths via the op amp supply terminals.

higher frequencies. Therefore, the simple low-pass filter shown in Figure 11-6a can be quite effective. Typically, R is on the order of 100Ω. Unfortunately, the operational amplifier d.c. supply currents must flow through R. The most general solution is to provide separate regulated supplies for the low-level and power-level circuits shown in Figure 11-6(b).

The example of Figure 11-5 is perhaps a worst-case situation, and it clearly illustrates the possible unwanted feedback paths which can occur due to limited PSRR. However, even single operational amplifiers are subject to the same effect if sufficient high-frequency decoupling is not provided between the operational amplifier supplies and ground. Of course, any inductance associated with the power supply line can only aggravate this problem.

Failure to account for limited PSRR is perhaps the basis for the common notion that if an operational amplifier is connected in a high-gain configuration, it is somehow less stable. Of course, considering only feedback paths between the operational amplifier and the conventional input terminals this idea is strictly untrue since *higher forward* gain means that *less feedback* is applied around the amplifier providing *greater stability*. However, as we have seen, a casual treatment of the operational amplifier supply terminals may create unconventional feedback paths whose adverse effects are emphasized by high forward gains.

We should stress that although power supply problems originate in a fashion similar to ground loop problems and may have similar effects they are *intrinsically* easier to solve. To appreciate this point let us reconsider the ground loop problem illustrated in Figure 11-2. To provide a viable communication link between System #1 and System #2 we needed to establish a common reference between two systems that previously had none. Similarly, in the simple grounding example where we provided individual ground lines for each circuit (Figure 11-1), our purpose was to create a good reference point between *two* separate locations. In contrast, the power supply decoupling problem is strictly local. We need only provide a certain level of performance at the supply terminals of a particular amplifier *relative* to its respective ground line. The relationship between the supply lines of different amplifiers is of little concern (some circuits, see Figure 11-5, are exceptions). Thus, we will easily satisfy the individual power supply requirements of Systems #1 and #2 in Figure 11-2 if we simply decouple the power supplies where they enter each system using techniques similar to those shown in Figure 11-6. How much more difficult it was establishing a single medium-frequency communication channel between the systems!

11.2.4   Mechanical Sources of Spurious Signals

Obvious sources of mechanically generated spurious signals include poor electrical connections and cold solder joints. Careful manufacturing procedures should eliminate this type of problem. There is, however, a very serious source of mechanically generated signals whose effect is not so easily eliminated, i.e., almost any multiconductor cable and especially coaxial cable. The effect may easily be understood by considering Figure 11-7(a). The resistive divider composed of R_1 and R_2 creates a d.c. bias on the center conductor of a shielded coaxial cable. The capacitance C of this cable is charged to the voltage V storing a charge Q given by:

$$Q = V \times C$$

Since the capacitance C is determined by the physical dimensions of the cable, if we deform the cable we will change its effective capacitance. Assuming we instantaneously apply a mechanical pressure that decreases the cable capacitance by ΔC, then the cable voltage will instantaneously increase to $V_1 = Q/(C - \Delta C)$ since the charge cannot immediately flow out through R_1 and R_2. Eventually, however, the excess charge will bleed out through R_1 and R_2 in parallel, returning the center conductor to its original voltage V.

Therefore, bumping or vibrating a voltage-carrying cable can cause a significant spurious signal. This effect will be more pronounced for higher

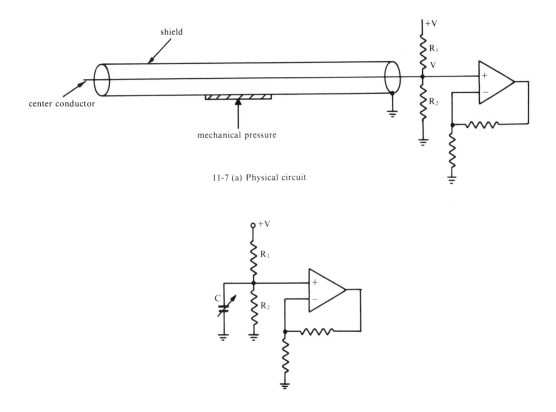

11-7 (a) Physical circuit

11-7 (b) electrical equivalent

Figure 11-7. Mechanical deformation of a signal carrying cable will induce an unwanted signal due to capacitance modulations.

circuit impedances, higher frequency mechanical disturbances and larger coaxial cables.

A differential signal carried in a shielded twisted pair cable will exhibit less of this effect than a coaxial cable. The only complete fix for this problem is to maintain critical cables immovable, but this may violate the original purpose for using most cables, that is, flexibility. However, adequate noise rejection in most applications may be obtained by moving a preamplifier stage to the sending end of the cable since the low impedance preamp output will absorb the charge pulses caused by cable deformations, with minimal effect on the voltage signal.

11.2.5   Reducing the Effects of Stray Fields

The effects discussed in the preceding sections could be explained by considering the currents that flow through the interconnections of our system. In the following sections we will explore the difficulties created within an electronic system by the presence of stray fields.

Three types of fields are of interest: the electric field, the magnetic field and the electromagnetic field. In the following sections, we will discuss the characteristics of each field and determine how to reduce their adverse effects. The electric field is treated first since it is responsible for the majority of field induced unwanted signals.

Eliminating the Effects of Electric Fields

Any a.c. voltage has associated with it an electric field. The effects of this field may be coupled into a circuit by stray capacitances. Consider the preamplifier of Figure 11-8 located on a laboratory bench. A three-wire, 115 volt, a.c. power line is distributed through the room. Recall that the green wire of the power line is referenced to local earth ground and is not used as a current-carrying conductor. The white wire is also referenced to local earth ground but since current flows in this line its potential may differ slightly from that of the green line at some points within the system. The black line carries a 115 volt a.c. signal. In accordance with conventional practice we have connected our circuit and chassis ground to the green wire, i.e., earth ground. One plate of the stray capacitance C_{S1} shown in Figure 11-8 is composed of the entire length of the black line and the other plate is formed by the length of wire connected between the 1 meg resistor and the operational amplifier noninverting input. Air is the dielectric medium. Now, the value of C_{S1} may typically be 0.05 pF and at 60 Hz its impedance is given by $1/(2\pi fC) = 5.3 \times 10^{10}$ ohms. With 115 volts a.c. on the black line, C_{S1} will therefore couple an unwanted signal to the noninverting input equal to

$$V_{ac} \times \frac{R_{in}}{X_c} = \frac{115 \times 1 \times 10^6}{5.3 \times 10^{10}} = 2.2 \text{ millivolts (rms).}*$$

Amplifier "a" may also be affected by stray capacitances to other sections of the system in which it is included. Consider the 20 kHz oscillator whose output line has stray capacitance C_{S2} to the input of amplifier "a". If the value of C_{S2} is 0.01 pf, its impedance at 20 kHz is $1/(2\pi fC) = 8 \times 10^8$ ohms, and the 10 volt rms output of the oscillator will cause a disturbance at the amplifier input equal to

$$V_{ac} \times \frac{R_{in}}{X_c} = 10 \text{ volt} \times \frac{10^6 M}{800 \times 10^6 M} = 12.5 \text{ m volts rms.}*$$

---

*In the denominator we omitted R_{in} with respect to X_c.

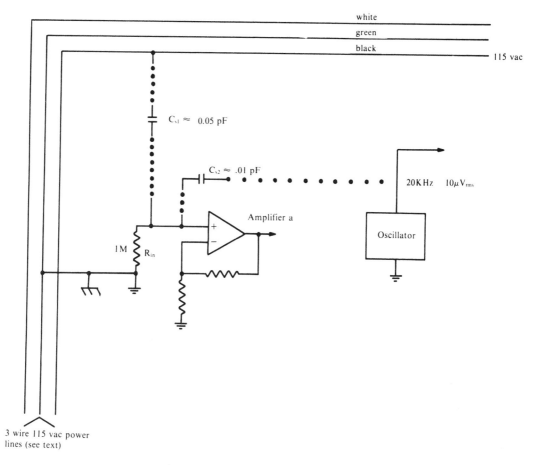

Figure 11-8. Electric fields may easily be coupled into a high input resistance stage by comparatively small stray capacitances.

Either of these signals is sufficient to degrade the performance of many systems.

We may avoid the effects of capacitively coupled signals by electrostatically shielding sensitive sections. Ideally, this shield would consist of a conductive box which totally encloses the area to be shielded. This box *must* be electrically connected to the local ground connection of the shielded area. As shown in Figure 11-9, effects of the stray capacitances are shorted to ground by the shield, allowing no internal fields. An effective means of routing the power supply leads into the box is also illustrated. In practice, an incomplete grounded shield may provide adequate signal rejection, depending upon its orientation. Of course, any portion of the circuit which is not shielded will be subject to the stray signals.

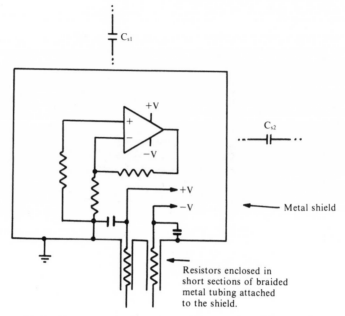

Figure 11-9. Unwanted signal currents associated with stray capacitances are shorted to ground by the shield.

Additionally, we should keep signal generators such as the oscillator of Figure 11-8 as far away from sensitive areas as possible, and within the sensitive areas keep impedance values to a minimum. As we have seen from the examples, the magnitude of the unwanted signal is directly proportional to the value of C_S, the frequency and amplitude of the interfering signal and the input impedance of the amplifier. Clearly, we should attempt to minimize any of these parameters that we can control.

Since the current *leads* the voltage in a capacitor, the voltage waveform of unwanted signals coupled into a circuit by stray capacitances will *lead* the voltage waveform that generates the unwanted signal by a $90°$ phase angle. This fact provides a convenient means of determining if an unwanted signal is being capacitively coupled into a circuit, assuming that we can isolate the unwanted signal source.

Eliminating Magnetically Induced Spurious Signals

Usually, magnetically induced signals are not a problem in low-level circuitry. Many times, however, stray magnetic fields are wrongly accused of causing unwanted effects. For example, the magnetic field associated with a power transformer is frequently blamed for spurious signal effects which are actually caused by the transformer winding *voltages* via stray capacitive paths.

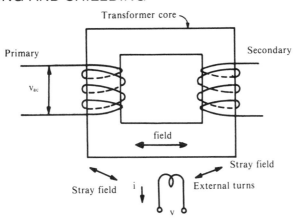

Figure 11-10. Analyzing the voltage and current wave forms created in an external loop by the stray magnetic field surrounding a power transformer.

To better understand those instances where magnetic fields *are* a problem, we may start by considering the stray magnetic field associated with a power transformer. In operation, the transformer of Figure 11-10 has an alternating magnetic field flowing in the core in the direction shown. With ideal coupling between the primary and the secondary, this field is determined solely by the primary voltage and core characteristics and is unaffected by any load current flowing in the secondary. However, some stray field will flow near the outside of the core and pass through the external turns as shown. If the turns are open circuited, then the signal v will be either in phase with the primary voltage or exactly 180° out of phase, depending on the polarity of the winding. Under these conditions the stray field in the vicinity of the external turns will be unaffected.

We normally associate a number with the primary winding referred to as the volts per turn, V_t, specified at the operating frequency. To determine the voltage output of another winding *on the core*, we multiply the number of turns of that winding by V_t since the same magnetic field flows through both windings. We may perform a similar calculation for our *external* winding, but we must multiply V_t by the *fraction* of the total magnetic field which passes through the external turns. For this multiplier to be as large as 0.001, the external turns must be quite large and close to the core. More typically, this multiplier will be on the order of 10^{-4} to 10^{-7} for smaller turns located several inches from the transformer core. For a medium power 60Hz transformer, V_t will be about 200 mv/turn and therefore, the induced voltage on a typical external open-circuited turn will range from 0.02μvolts to 20μvolts. Contrast this signal level with the millivolt disturbances caused by capacitive coupling discussed earlier.

Now, if the external turn is short circuited, a current *i* will flow in the wire that will effectively cancel the field in the vicinity of the turn but, since the external turn is so loosely coupled to the transformer field, this will have little effect on the primary waveforms. In a well-designed 60 Hz power transformer *of any size* the magnetic intensity, H, will have a typical peak value of 10 amps/meter inside the core and near its surface. Thus, the induced current in a 1 cm dia. shorted external turn placed near the core material will be about 10 amp/meter × 0.001 meter = 10 mA. This current will be in phase with the magnetizing current and therefore 90° out of phase with the primary voltage. The size of the power transformer will affect this induced current to the extent that the field H will fall off more slowly around a larger transformer as compared to a smaller transformer. Otherwise, the transformer size has *no effect* on the magnitude of induced magnetic signals!

From the preceding discussion, we see that the same shielded box which suppressed capacitvely induced unwanted signals, will suppress unwanted magnetic effects only if a current-carrying loop exists within the box since this ensures that all internal magnetic fields will be canceled by the surface currents.

Perhaps the best overall solution is to enclose all critical transformers in grounded conductive containers, thereby eliminating both the electric and the magnetic external fields.

Eliminating the Effect of Electro-Magnetic Fields

An alternating electric field always has associated with it an alternating magnetic field and vice versa. However, at low frequencies, we may easily separate the effects of these fields and analyze them independently. Thus, we were able to treat the magnetic field in the core of a power transformer separate from the electric field in the primary circuit which causes it. However, at higher frequencies, the electric and magnetic fields may become associated with each other in free space as a traveling electromagnetic wave such as radio and television signals. Under these conditions, the effects of the electric and magnetic waves are inseparable. Controlling these waves is particularly difficult since they represent *energy* traveling through free space. Recall that the previously discussed solutions to our electric and magnetic field problems resulted in circuit paths that were imaginary and therefore involved no energy loss. However, we cannot simply *short out* an incident electromagnetic wave since, as would be the case with a shorted transmission line, the energy will simply be reflected and could cause further problems.

If we are to reduce the effects of electromagnetic waves, we have two choices: 1) reflect or reroute the energy to an uncritical area. Due to the impedance mismatch between a conductor and free space, sheets of copper may be used as an effective reflector; or 2) absorb and dissipate the incident energy. The ability of a material to stop an electromagnetic wave is measured by its *skin depth*. This number represents the depth at which the incident field falls to $1/e$ of its value at the surface where e is 2.7183 . . . the base for the natural logarithms. Mathematically,

$$\delta = \sqrt{\frac{2\rho}{\omega\mu}} \qquad \textbf{(11-1)}$$

where δ is the resistivity of the material in ohm-meters, ω is the angular frequency ($2\pi f$), $\mu = \mu_0 (1 + X_m)$* and δ is the skin depth in meters. The electric field is attenuated exponentially as the wave travels through the material and for maximum effectiveness the material should be several skin depths thick. To minimize the skin depth, we need to maximize μ and minimize ρ. Many materials, such as Mumetal, have these specifications and are useful in these applications. Only circuits which are surrounded by these materials will be protected. Some of the incident energy will reflect from the surface of such a shield and could cause problems in other parts of a system.

In many applications, a layer of copper is bonded with a layer of reasonably conductive high μ material providing both a reflecting and absorbing surface. In general, low-frequency electromagnetic waves are harder to reflect or absorb than higher-frequency waves.

11.3 ELIMINATING SPURIOUS SIGNALS IN DIGITAL CIRCUITS

All digital circuits are designed to withstand a certain level of unwanted signal voltages on the interconnection lines without degrading device performance. This voltage, V_{NI}, is called the *noise immunity* of the logic gate (see Chapter 8) and is largely determined by gate *output* voltage levels *and* gate *input* voltage requirements. For example, the minimum guaranteed output *high* voltage for a low-power Schottky TTL gate under maximum loading conditions is 2.7 volts, but the guaranteed input logical high voltage for the same gate *input* is 2.0 volts yielding a "noise" immunity of $V_{NI} = 2.7 - 2.0 = 700$ mV in the output high state. Thus, –700 mV peak of unwanted signals could exist on an interconnection between two gates of this logic family without upsetting circuit operation, assuming that the driving gate is in the high state. The "noise" immunity of this logic family in the low state may differ from this value as will the "noise" immunity of other logic families. However, we can always calculate the appropriate V_{NI} from the manufacturer's published data. The real problem for the circuit designer

---

*$\mu_0 = 4 \pi 10^{-7}$ H/m, X_m is the magnetic susceptibility and is dimensionless. $1 + X_m = M_r$ = relative permeability.

is determining the actual unwanted signal levels, v_x, that will occur in this final design and ensure that the peak values of v_x are smaller than V_{NI}. In the following sections we will describe the major unwanted signal sources and the factors that determine the effects these sources have on the interconnections of various logic families.

Table 11-1 is a listing of the performance characteristics which most directly relate to the production and reception of unwanted signal voltages in digital systems. The four most popular logic families are represented.

	Schottky TTL (STTL)	Low Power Schottky TTL (LS TTL)	CMOS 4000 Series 5 volt supply	Emitter Coupled logic 10,000 Series (ECL)
Output Voltage swing	$\approx$3V	$\approx$3V	5V	800 mV
Rise time	3ns	10ns	90ns	2ns
Fall time	2.5ns	6ns	100ns	2ns
Edge speed	$\approx$1V/ns	.35 volt/ns	.05 volt/ns	.3 volts/ns
Power supply current pulse during switching for unloaded gate	$\approx$30 mA	8 mA	<1 mA to +5 mA (very dependent on input rise and fall time)	<1mA
Maximum interconnection length for open lines	12 in.	12 in. (see text)	no limit due to high output resistance (see text)	8 in.
Maximum Flip flop toggle rate	125 MHz	45 MHz	3 MHz	150 MHz
Output resistance in high state	53Ω	150Ω	$\approx$2 kΩ	7Ω
Output Resistance in low state	7Ω	25Ω	$\approx$1 kΩ	7Ω.
Input resistance & capacitance	2.8kΩ & 3-5 pF	20kΩ & 5-7 pF	4 pF	50 kΩ & 3 pF

Table 11-1. Important characteristics of Digital Gates relative to production and reception of spurious signals.

Note that we have chosen to list CMOS characteristics at a 5-volt supply level to allow evaluation of CMOS logic used in conjunction with other 5-volt logic families. However, when operated at higher voltages, all of these specifications change in a direction that reflects a significant improvement in overall performance. We will refer repeatedly to Table 11-1 in the next sections.

11.3.1 Unwanted Signal Coupling via Supply Voltage Connections

In TTL and CMOS devices, the output high and low voltage levels vary almost directly with any signal which appears on the +5V supply lines and ground respectively. Thus, a 300mV drop in the +5V supply line will cause a drop in the output high voltage level only slightly less than 300 mv. Similarly, signals on the ground line affect the output low-level voltage. As was the case with analog circuits, currents flowing in the resistance and inductances associated with the power distribution's network will cause unwanted potential differences within the system. These currents can be surprisingly high, creating transient voltage differences on the ground or supply line that may exceed the "noise" immunity of the logic family. This would cause erroneous circuit operation due to the output voltage dependencies described above.

One portion of this transient current is composed of the charge delivered to any load capacitance when the output of a digital gate switches state. Figure 11-11 shows the path of current flow in the output stage of an LSTTL gate under these conditions. The value of the charging current may easily be determined if we know the output transition edge speed (slew rate) in volts per nanosecond since during the transition,

$$I = \frac{\Delta V}{\Delta t} \times C \qquad (11\text{-}2)$$

if C is expressed in nanofarads.

In a typical system a LSTTL gate may drive 10 LSTTL inputs creating a load capacitance of about 60pF (see table 11-1). Now, the edge speed of LSTTL is given by table 11-1 as 0.35 volts/ns. Thus,

$$I = 0.35 \frac{volts}{ns} \times 0.060 \ nF = 21 \ mA.$$

Under the same loading conditions, a STTL gate would create a transient current equal to 60 mA. The duration of both of these transients is equal to the rise time of the driving gate as shown in Figure 11-11.

There is an additional transient current present even when the gate

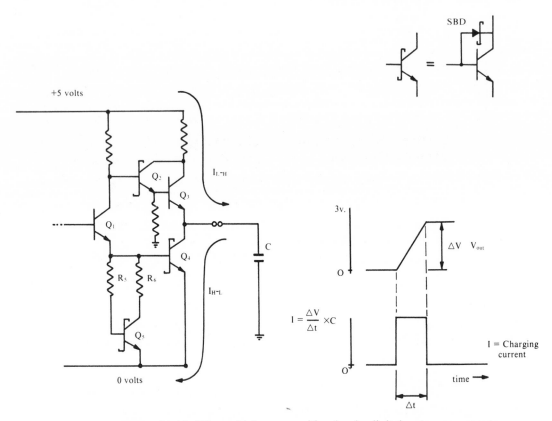

Figure 11-11. When driving capacitive loads digital gates can create transient currents which must flow in the supply lines if improperly decoupled.

output is unloaded, caused by simultaneous conduction of the output transistor Q_3 and Q_4 (Figure 11-11) during an input transition. This current is especially bothersome during the low-to-high output transition in TTL gates since Q_4 cannot turn off as fast as Q_3 turns on although transistor Q_5 (which was not present on pre-Schottky TTL circuits) acts to help equalize the switching times of Q_4 and Q_5. Typical values of this current transient are listed for various logic families in Table 11-1. This effect is also present in CMOS devices, but the value of the current transient is not easily predicted since the output resistance of a CMOS gate varies widely.

ECL logic circuits are less prone to the generation of power supply signals *and* less sensitive to any signals which do exist. Since the edge rate of the ECL output waveform is smaller than that of either STTL or LSTTL, the currents required to charge and discharge load capacitances are smaller. Furthermore, the ECL output voltage swing is about three times smaller

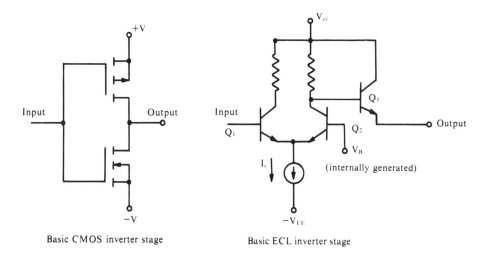

Figure 11-12. Basic CMOS and ECL inverter stages.

than TTL's and therefore, the total *energy* moved into or out of the load capacitance is nine times less. In addition, the balanced differential amplifier structure of the basic ECL inverter stage (Figure 11-12) cannot generate a current spike if Q_1 and Q_2 conduct simultaneously since the total current through the pair is I_c under all conditions. Now, of course there will be a change in current flow through Q_3 since the outputs of ECL gates are connected externally to a pulldown resistor. However, this current change has the same waveshape as the output voltage change. This contrasts with the current waveshape associated with charging load capacitances which reaches the value calculated from Equation 11-2 when the voltage transition first begins as shown in Figure 11-11. This waveshape clearly contains more disturbing high-frequency components than the output voltage waveshape.

Finally, the structure of the basic ECL gate results in significant rejection of unwanted signals on the negative supply, V_{EE}, allowing the designer some leeway in his P.C. board layout of the negative supply. For this reason, the positive supply of ECL circuits is usually connected to circuit "ground," the best referenced potential in most designs. There is no magic about this circuit connection, but if the V_{EE} supply is grounded, then the designer must exercise care in the layout of the positive supply V_{CC} to minimize V_{CC} voltage differences between packages.

The solution to any of the supply current transient problems described above is to provide a low impedance path between the positive and negative supply terminals of the logic gates. This is normally done by decoupling the supplies with a capacitor. The particular type and structure of this capacitor is, however, very critical. The oft-heard expression "decouple the power supply lines with good quality r-f capacitors" is at best misleading and could easily be branded simply incorrect. In fact, the most advantageous type of decoupling capacitor is a ceramic monolithic structure with a Z5U or other high K ceramic dielectric.* Now, a look at Figure 2-7 (Chapter 2) will convince anyone that high K ceramic dielectric is not "good quality r-f-material," but several factors combine to yield the unique utility of this capacitor. First, the monolithic construction and high K material results in a low inductance structure with maximum capacitance per volume. The fact that the capacitance decreases with frequency is not a serious limitation—there is still plenty of capacitance at high frequencies. Second, the increased dissipation factor at high frequencies actually helps damp-out any resonant effects on the supply line and is, therfore, advantageous. In contrast, a *high-valued* capacitor constructed from good quality rf ceramic material (low K) would have a very significant inductance (due to increased size requirements) that would cause self-resonance at a frequency of several mega Hertz. This resonant frequency is quite disturbing in any digital system.

Since *load* capacitors (those responsible for generating transients) may be located some distance from the driving gate, transient currents will flow between devices even if each device is properly decoupled. For this reason, the interconnecting supply lines must themselves be low impedance paths. Ideally, we should provide two complete parallel copper planes on PC boards to be used as positive and negative supplies, since this would provide the lowest possible supply line impedance between the separate devices on the PC board. Of course, if we are using two-sided PC boards, this would leave no room for interconnecting signal paths. As a practical solution when using the high-frequency logic families, e.g., STLL and ECL a* large portion (50% - 70%) of one side of the PC board should be devoted to a ground plane. As mentioned, no amount of supply decoupling will adequately compensate for a poorly conceived power distribution system, but a good supply layout can reduce the total number of supply decoupling capacitors required in a given system. The recent introduction of power supply bus bars for PC boards† has provided the designer with an excellent solution to high impedance power distribution problems. In addition, their use largely frees both sides of the PC board for signal connections, thereby shrinking the overall board size and further reducing high frequency effects.

It should be noted that the ideal power supply structure

---

*Typical example: USCC/Centralab BME - radial CR series monolithic capacitor 25 volt 0.1 μf.
†Available from the Rodgers Corporation and others.

recommended for digital circuits completely violates the precepts of analog circuit layout described earlier. In analog design, the ground plane recommended for proper digital layouts would behave as an infinite number of ground loops. This apparent contradiction can be resolved by recognizing that 1) a digital circuit has a level of "noise" immunity which can tolerate some supply and ground line spurious signals, and that 2) the frequencies normally encountered in analog design do not require the transmission line environment necessary for high-speed digital systems.

11.3.2 Controlling Capacitive Effects Between Signal Lines

Figure 11-13 illustrates the origin of the capacitive coupling effect between signal lines. When gate #1 switches high, the stray capacitance between lines l_1 and l_2 will couple a signal onto line l_2. If the edge rate of the disturbing signal were extremely high, then the peak value of the coupled signal could be as large as the voltage transition of line l_1. Such a worst-case situation will rarely occur, but we may say that the magnitude of the coupled

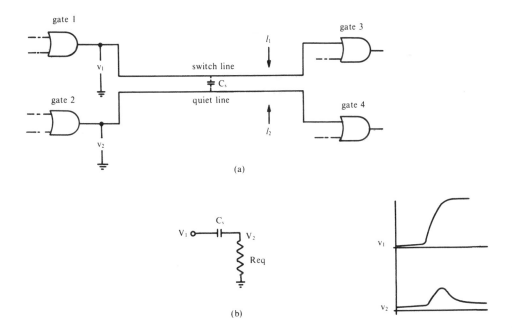

Figure 11-13. Capacitive coupled signals may affect adjacent lines.

pulse is directly proportional to edge speed, the value of C_S, and the value of R_{eq}. (Although not considered in this discussion, any capacitance on l_2 with respect to ground will reduce this effect.) Now, the edge speed is determined by the logic family in use; the stray capacitance, C_S, is determined by the size, proximity and length of l_1 and l_2 (and may easily exceed 20 pF for closely spaced lines on glass epoxy PC boards); and, since the input resistance of most logic families is much higher than the output resistance, we may say that R_{eq} will be determined by the output resistance of the logic family in question.

Under adverse conditions, the value of v_2 can exceed the noise margin of gate 4 and thereby propagate a spurious signal. Clearly, the way to limit this unwanted pulse is to minimize the three factors described above, i.e., edge speed, C_S, and R_S. For the most part, the designer must get a practical feel for the magnitude of v_2 under various circumstances and with various logic families on operating PC boards, since a precise determination of v_2 prior to breadboarding is quite difficult. However, the following suggestions and warnings can be useful in avoiding some problems and solving those which were not anticipated.

1) If one logic family is used throughout the design, problems will only occur between parallel lines that are closely spaced and travel some distance together.

2) Beware of open collector TTL gates since the required pull-up resistor will determine the output resistance in the high state, and this value will typically be 5-10 times larger than that shown in Table 11-1.

3) Run critical lines perpendicular to disturbing lines since this limits C_S.

4) The stray capacitance between two lines on a PC board will be up to 5 times larger than the capacitance between the same lines in air since the dielectric constant of the PC material is greater than 1. A board material change could also cause similar problems.

5) Serious problems arise when logic families are mixed since high output resistance circuits are especially prone to capacitively coupled signals. Of special note is the high value for CMOS gates operated from 5 volt supplies. Such high output resistances make the much-advertised high noise immunity of CMOS gates of little value when used in conjunction with TTL.

11.3.3   Sweeping Signal Reflection Problems Under the Rug

The interconnecting lines of any circuit board will act like transmission lines at high frequencies. Therefore, if proper line matching procedures are not followed, unwanted reflections can degrade our signals and result in spurious system operation. Since the subject of transmission theory is quite involved and the practical application of this theory is only slightly less involved, we should like to avoid worrying about these problems as much as possible. Fortunately, a very simple rule and even simpler trick will allow the designer to bridge these problems with surprisingly few compromises.

The Very Simple Rule

Unless the very simple trick is used (see next section), we must keep signal interconnection lengths, l, such that

$$l_{max} \leq \frac{t_r}{2\,t_{pd}} \qquad (11\text{-}3)$$

where t_r is the rise time of the gate output in nanoseconds and t_{pd} is the propagation delay of the transmission line in nanoseconds/inch and l is in inches.

A t_{pd} of 0.150 ns/in is typical for a PC board conductor over a ground plane. Exceeding this limit may result in a reflection which is larger than the noise margin of the logic family. Recommended maximum line lengths are shown in Table 11-1. Note, however, that the value shown for LSTTL is much shorter than expected from application of Equation 11-3. This results from a peculiarity of the input structure used in LSTTL as supplied by some manufacturers. If excessive current flows in the Schottky diode connected between the input and the ground terminal, internal parasitic transistor action can affect the state of the output stage, causing the device to deviate from its truth table. Keeping lead lengths below 12 inches will minimize the current flow in this clamp diode by reducing negative overshoots following the positive-to-negative transition of the driving gate. (Consult the manufacturer of your selected device for information relating to this problem.) CMOS has an unrestricted line length for reasons discussed below.

The Simple Trick-Series Damping

A resistor placed in series with the output of the driving gate and located as close as possible to the gate itself will allow *any logic* family to drive *any line* length, see Figure 11-14. The value of this resistor should be $\approx$ 24 ohms although many systems will work with smaller values and some will require larger values. Since the transmission line will appear resistive over a range of frequencies determined by its length, R_S in series with the gate output limits energy injected into the line at the critical high frequencies. We must, however, compromise rise time at the receiving end since it will be slowed depending on the line length. This is usually not a limitation since we can frequently group the high speed sections of a complex system in small areas within which transmission line effects are negligible. If, however, we

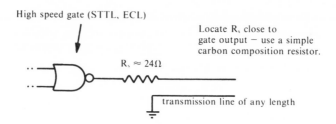

Figure 11-4. Using series damping to allow signal transmission over lines of any length.

wish to create a high-speed data communications channel, series damping is not applicable. Use of series damping will limit fan-out of TTL gates since d.c. currents must flow through R_S. In general, lines which travel over ground planes are less trouble than those which do not. We see immediately that CMOS has no line length limit because its output resistance is rather high.

Please note, that these solutions do not eliminate reflections; they simply limit them to values which are tolerable in most applications. To completely eliminate the reflections, we must properly terminate the transmission lines. A great deal of data on this subject is available from the manufacturers of the high-speed logic families, especially ECL manufacturers.

11.4   THE HYBRID DIGITAL - ANALOG SYSTEM

As we have seen the prescriptions for effective control of unwanted signals in digital and analog circuits are in basic conflict and so the best advice to the hybrid system designer is to keep the digital and analog sections as independent from one another as possible. Within their respective areas, use the appropriate spurious signal control techniques outlined in this chapter. However, some of the suggestions listed below may prove helpful.

1)   Connect the digital and analog grounds together at only one point.

2)   You must provide separate supply voltages for digital and analog sections.

3)   The high edge rates of digital gate outputs may easily couple capacitively into the analog circuitry. Use electrostatic shielding around the analog circuit connected to *analog* ground potential.

4) Usually, the number of interconnections between the analog and the digital system are few and the designer should consider electro-optic coupling as an inexpensive means to increase circuit isolation.

5) If analog circuitry must be used in conjunction with very fast digital circuitry, e.g., fast analog-to-digital conversion, the designer should strongly consider using Emitter Coupled Logic, ECL, exclusively in the digital section instead of STTL or LSTTL. ECL's lower edge rates, lower signal level, and freedom from power supply transient currents offer significantly lower levels of spurious signal generation in spite of its high operating speed capabilities.

Chapter 12

System Design

12.1 In the previous chapters we discussed using I.C.'s to perform specific functions such as full wave rectification or analog-to-digital conversion. Usually, there are several ways to realize a given function and the designer must choose among them based on *system* requirements. For example, when choosing an operational amplifier for use in a functional block, e.g., an oscillator, the designer must select a part which has the appropriate bias current, bandwidth, etc. to ensure successful operation. Similarly, the *system* designer must choose the individual functional blocks so that system requirements are met without increasing costs due to overspecification.

In this chapter we shall consider two systems: one analog and one digital. First, we shall take an overview of each problem and rough out some block diagrams. Then, since the system designer is usually responsible for the entire electrical design, we shall "fill in" the blocks using the results of previous chapters. Both of the systems represent current design practices. The analog system is part of an instrument used to characterize nasal breathing and is currently in production. The digital system is a part of a transient recorder developed under contract at the University of Pittsburgh.

12.2 DESIGNING ANALOG SYSTEMS

Many analog systems have as their input a transducer of some kind and the system is expected to provide an output which is the *analog* (voltage or current) of the variable which the transducer is monitoring. For example, a thermistor is a two-terminal device whose resistance is a function of temperature, and when used in a temperature indicator the associated analog circuitry would measure these resistance changes and provide an output voltage which is proportional to the thermistor temperature.

In our system design example we shall consider the circuitry required to bias a pressure transducer, measure the transducer output, and provide a voltage analog of the real-time pressure.

12.2.1 Interfacing a Differential Variable Reluctance Transducer (DVRT)

Many systems use a DVRT as a sensor for position displacement. Briefly, the DVRT consists of two inductors L_1, L_2 (Figure 12-1) coupled by a magnetic part which is responsive to a mechnical stimulus. Motion of the mechanical part in one direction causes the value of L_1 to increase and L_2 to decrease and vice versa for motion in the opposite direction.

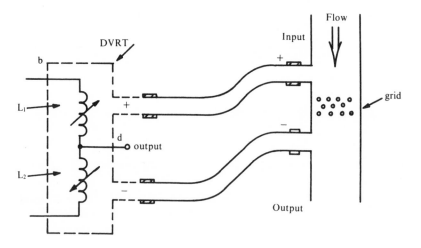

Figure 12-1. Differential variable reluctance transducer (DVRT) and flow sensor.

In a pressure transducer the mechanical part is a thin stainless steel (magnetic) diaphragm which separates two chambers. When a pressure differential exists between the chambers, the diaphragm is deflected, causing two inductances to vary in opposite directions.

In this sample design, we shall develop a system that monitors the flow associated with the inspiration and expiration of human breathing. The most accurate means of sensing and measuring flows of this small magnitude utilize a DVRT configured as a pressure sensor in conjunction with a flow sensor as shown in Figure 12-1. All air flow is channeled through the flow sensor where it must pass through a grid, causing a pressure differential which is then applied to the DVRT. To avoid affecting the breathing pattern of the patient, the pressure drop in the flow sensor must be kept small, typically on the order of 5 mm H_2O.

To sense the inductive (reluctive) change we must provide a.c. bias signals to the transducer as shown in Figure 12-2. Signals b and c, which are sine waves symmetrical about zero volts, but 180° out of phase, are applied to the DVRT. If $L_1 = L_2$, then the signal at d will be zero. However, if $L_1 < L_2$, d will be a sine wave whose magnitude is determined by the magnitudes of a and b and the difference between L_1 and L_2, and the output will be in phase with b as shown. When $L_2 < L_1$, the output magnitude varies as before, but now, d is in phase with c. Control signal e is high when b > 0 and low when b < 0 and if we observe d only when e is high, we see the shaded portions of the graphs of d for positive and negative pressures .Viewed in this fashion, we see that for positive pressures the output d is a *positive* half-wave signal and for negative pressures d is a *negative* half-wave signal whose magnitude is directly proportional to the difference between L_1 and L_2 for small changes in L_1 and L_2. If control signal e operates a switch which connects d to the input of a low-pass filter when e is high, then the a.c. components of the half-wave input will be filtered out leaving a d.c. signal which is proportional in magnitude and equal in sign to the original air flow.

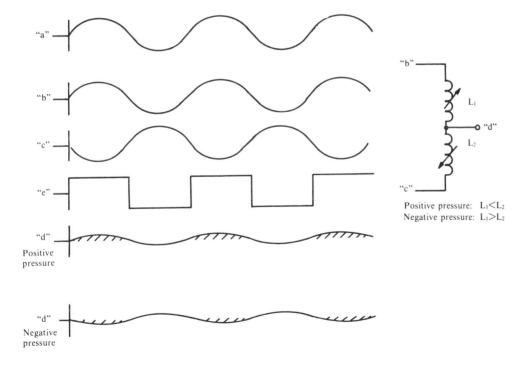

Positive pressure: $L_1 < L_2$
Negative pressure: $L_1 > L_2$

Figure 12-2. Required biasing signals for the DVRT.

Figure 12-3 shows a block diagram of the measurement system with some detail in the transducer area. In Figure 12-4 a circuit diagram of the completed system is shown. Working from left to right, we shall develop the circuitries which are represented in the blocks of Figure 12-3.

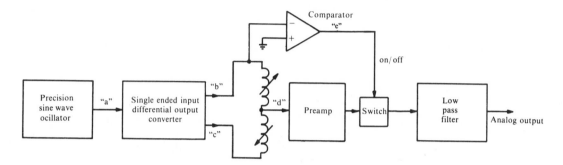

Figure 12-3. Hybrid block diagram of flow measuring system.

Precision Sine Wave Oscillator

As we have mentioned, the transducer output voltage for specific unequal values of L_1 and L_2 is proportional to the bias voltage a and b. Therefore, to achieve stable operation, we must maintain tight control of the voltage amplitude at points a and b and this is achieved by controlling the output of the precision sine wave oscillator. This oscillator, shown in Figure 12-4, is almost identical to the oscillator of Figure 4-10 (Chapter 4) except for the addition of operational amplifier #3 which is operated as a precision half-wave rectifier (Chapter 6). In the original circuit of Figure 4-10, D_1, operated as a half-wave rectifier, was used to sense the amplitude of the oscillator output. Although this sensing technique is satisfactory for many purposes, the temperature sensitivity of the diode forward voltage drop results in some amplitude variation over the operating temperature range. By using a precision half-wave rectifier in place of D_1 these temperature dependencies are eliminated.

The Wien bridge components have been chosen for oscillations at 6 kHz. Unity gain compensated operational amplifiers will perform well in this section.

Single Ended Input-Differential Output Converter

Although a differential signal could be created by simply inverting the output of the oscillator with an operational amplifier, the differential transformer shown in Figure 12-4 was chosen for this purpose for the following reasons:

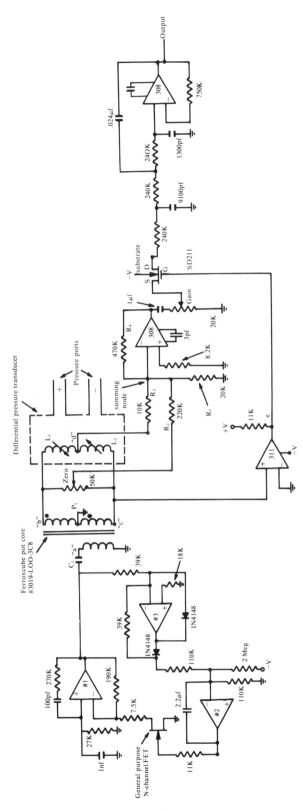

Figure 12-4. Circuit diagram of the flow measuring system.

a) The voltage magnitudes at b and c can be set equal by simply using equal numbers of turns in both secondary windings.
b) Since c is exactly the inverse of b, harmonic distortion products present at a have little effect on circuit performance.
c) Offset voltage problems and their temperature dependencies are eliminated.
d) At 6kHz, L_1 and L_2 represent a reactance of $\approx 750\Omega$. The transformer can be used to raise this impedance so that special buffer circuits are not required at the output of the oscillator.

To design the differential transformer, we must first choose a turns ratio. For the Validyne DP45 transducer used in this example $L_1 = L_2 = 20$ mH and therefore the reactance at 6kH of L_1 and L_2 in series $2 \times 2\pi \times$ 6kHz $\times$ 20 mH= 1500 KΩ. To utilize the full voltage swing of the oscillator we should transform this value up to about 3kΩ and therefore $N_p/N_s = \sqrt{3000\Omega/1500\Omega} = 1.4$, where N_P and N_S are the number of turns in the primary and secondary windings respectively. Note that N_S must be divided equally between the two secondary windings. To avoid further reactive loading at the primary, we make the primary reactance X_P of the transformer somewhat larger than the reactance reflected into the primary from the secondary. Let the primary reactance $X_p = 3 \times 3$ kΩ = 9 kΩ. Therefore, the primary inductance L_p must be $L_p = X_p/2\pi f = .4$ Henry. We find that this value of L_p can be obtained using an ungapped Ferroxcube 3019 Port core in 3C8 material with 220 primary turns. The secondary will contain $220/1.4 = 157$ turns divided equally between the two windings. Using #30 wire for the primary and #28 for the secondary, these windings will easily fit the 3019 core. The total reactance seen by the oscillator output is 3 KΩ $\|$ 9 KΩ=2.25 KΩ. Although this value seems rather low for an operational amplifier load, remember that this is a purely inductive load and therefore maximum current flows at zero volts where the operational amplifier's ability to supply current is much greater than it would be near the supply voltages.

Preamp and Zero Adjustment

Except for the presence of resistor R_1 (see below) the pre-amp stage is just an inverting amplifier. The gain for inputs to resistor R_3 is R_4/R_3=47. By summing a signal from potentiometer P_1 with the output of the pressure transducer, we can cancel any differences between L_1 and L_2 and thereby establish a precise zero output when there is zero pressure differential. In addition, this pot will allow us to offset the analog output to any d.c. value; a useful feature in some pressure detection applications. Since the voltage signal at the wiper of P_1 can be quite large, we choose R_2 so that the pre-amp gain from the zero pot input is about 2. Only a.c. signals pass through the pre-amp and therefore, we use a.c. coupling from the output of the pre-amp to avoid any offset errors.

Of special interest is the function of R_1. The LM308 operational amplifier should not be unity-gain compensated since it must provide accurate reproduction of a 6 kHz signal with a closed-loop gain of 47. With a 3 pF compensation capacitor, the 308 will be stable for closed-loop gains greater than 10 so it appears that there would be no problems. However, for frequencies in the megahertz range the impedance looking back into the pressure transducer becomes very high and the closed-loop gain will be determined by R_2 and R_3 or about 2. Without R_1, the circuit would oscillate in the megahertz range. With R_1 in the circuit, the closed-loop gain can never be less than about 23 and the circuit is quite stable. Note that at 6 kHz, the impedance at the summing node will be less than 100Ω due to the high open-loop gain of the operational amplifier and therefore, the addition of R_1 will not affect the 6 kHz performance of this amplifier.

Comparator and Switch

As shown in Figure 12-4 a 311 type voltage comparator is used to provide control signal e. Since the preamplifier inverts the transducer output signal we set e in phase with c instead of b (as shown in Figure 12-2) so that positive pressures cause positive output voltages. The Signetic SD 211 will be *on* when the gate is about 3 volts more positive than the source and it will be *off* when the gate voltage is equal to or less than the source voltage. With the open emitter tied to -15V the gate signal is sufficient to allow analog switching of $\pm$ 10 volt signals. Because the signal at c passes through zero so quickly, hysteresis is not required around the 311.

Low-Pass Filter

The input to the low-pass filter is an amplitude-modulated 6kHz half-wave rectified sine wave. This filter must reject 6kHz and higher harmonics while passing the lower frequency modulation. Therefore, to ease the filter requirements the maximum modulation frequency $f_{m(max)}$, should be much less than the carrier frequency f_c. In our design, f_m is the frequency of pressure variations and f_c is 6 kHz. A d.c. to 100 Hz frequency response is common in medical instrumentation and so we let $f_{m(max)} = 100$Hz. The three-pole filter shown in Figure 12-4 is unity gain in the low frequency passband with a 3 dB down frequency of 100Hz (Chapter 7). With three poles the negative slope in the stopband is 18dB/octave and therefore the attenuation at 6kHz is greater than 80 dB. The 3 dB down point of this filter may easily be changed by a linear scaling of either the resistors or the capacitors. For example, to move the 3 dB down point from 100Hz to 1000 Hz, simply *divide* all capacitor values *or* all resistor values by 10.

In some applications, $f_{m(max)}$ will be much closer to f_c than in our example and the three-pole filter may offer inadequate carrier rejection. If this is the case, simply cascade the three-pole filter to achieve 6 or 9-pole responses.

12.3 DIGITAL SYSTEM DESIGN

As before, we begin a digital system design with a block diagram showing the relationship between simple blocks within the system. In any well-designed digital system, operation is synchronized to a master clock. Various gates, flip-flops and MSI functions within the system will determine *what* the digital signals are and *where* they will go, but the master clock determines *when* these events will occur. A synchronous digital system can be viewed as a sequence of digital states. The system generates a new set of data states following the clock transition based on the data states prior to the transition. Thus, in addition to the block diagram, the digital designer must establish a *timing diagram* which shows the desired sequence of states mapped as a function of clock transitions. In the following sample design we shall discuss a typical timing diagram in addition to developing a block diagram and its associated circuitry.

12.3.1 Designing a Digital Date Recorder

In this design we shall consider a special purpose date recorder. The recorder is a battery-operated unit which is placed in a field location and visited once a year. Following field installation the recorder must keep track of time with a resolution of 1 second and upon reception of an "event occurred" pulse from a companion instrument, the recorder must enter the time of occurrence into a memory. The recorder must be able to store 30 such records. A year later, the recorder is addressed and the sequence of dates externally correlated with information contained in the companion instrument. Since the instrument is unattended for a one-year period, we must use an all CMOS digital I.C. design approach to achieve the desired battery life. Figure 12-5 shows a very simplified block diagram illustrating the major system blocks and the input lines. The binary counter is clocked at the rate of 1 Hz by a crystal clock. Thus, the LSB of the counter is incrementing at a 1-second rate. With 25 stages, the counter can record $2^{25} = 33,554,432$ seconds, which is somewhat longer than one year. At first glance, it might seem better to use digital clock circuits which are commercially available. Unfortunately, these circuits record the time in a BCD format and would require 50 bits to generate the month, day, hour, minute and second. This would double the size of the memory which is the most expensive single part in the system.

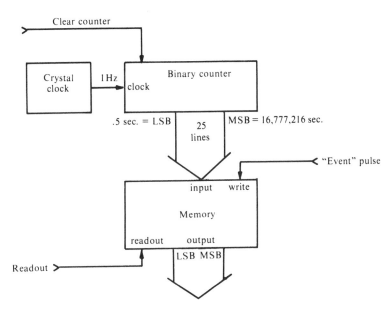

Figure 12-5. Simplified block diagram of date recorder.

Several aspects of our simplified block diagram are not practical:

a) A memory with 25 parallel input lines will require several packages and would not be cost effective. Since a single 1024-bit (1k) Random Access Memory RAM could hold all of the data we hope to collect and is available in a 16-pin dual-in-line package we will design our system around this type of memory.

b) Since the event pulse is independent of the crystal clock it is possible that the event pulse could occur while the binary counter is incrementing. This could cause erroneous data to be written into the memory. Therefore, we must separately record the arrival of an event pulse and send a write command to the memory only when the counter output data is stable.

An improved block diagram is shown in Figure 12-6. A crystal oscillator operating at 32.768 kHz drives a 15-stage frequency divider providing an output of 1 Hz since $2^{15}=32768$. The time information is stored in a 25-bit binary counter which is continuously clocked at the 1 Hz rate. The 25 parallel output lines of this counter are connected to the 25 parallel input lines of a parallel-to-serial (P-S) converter. Initially, the P-S converter is held in the parallel entry mode and the RAM address counter is reset. When an event pulse occurs the Control Circuit (CC) notes this occurrence and then, by observation of the 25-bit counter clock line, the CC chooses a time when the counter output lines are stable. At this time, the P/S control line of the P-S converter is taken low, locking the parallel data into the converter in

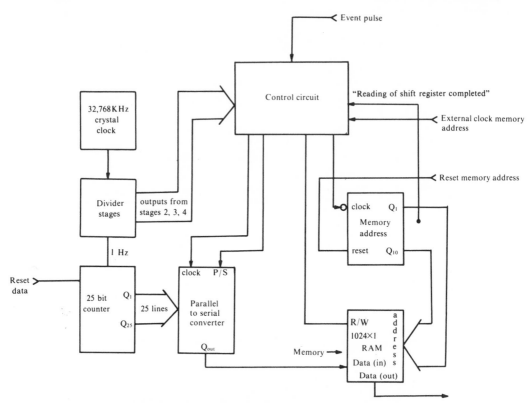

Figure 12-6. Detailed block diagram of Date Recorder.

preparation for serial readout to the RAM. The RAM is then placed in the write mode and the CC begins simultaneous clocking of the P-S converter and the RAM address. Thus, as the RAM Address increments 0, 1, 2, 3, etc., each successive bit of serial information is written into successive locations in the RAM. By analyzing the outputs of the RAM address the CC can determine when all 25 bits have been loaded into the RAM. At this time the system returns to its initial condition, except that the RAM Address counter is left at its last count.

Figure 12-7 shows a complete circuit diagram of the date recorder and Figure 12-8 is the timing diagram for this circuit. In this design, 7 auxiliary bits have been added to the 25 time bits. Thus, a total of 32 bits will be read into the memory upon the arrival of an event pulse. These additional bits could be used to store information about the ambient temperature, humidity, etc. at the time of the event. As we shall see, it will be easy to determine the end of the shift sequence since 32 is a power of 2. We will

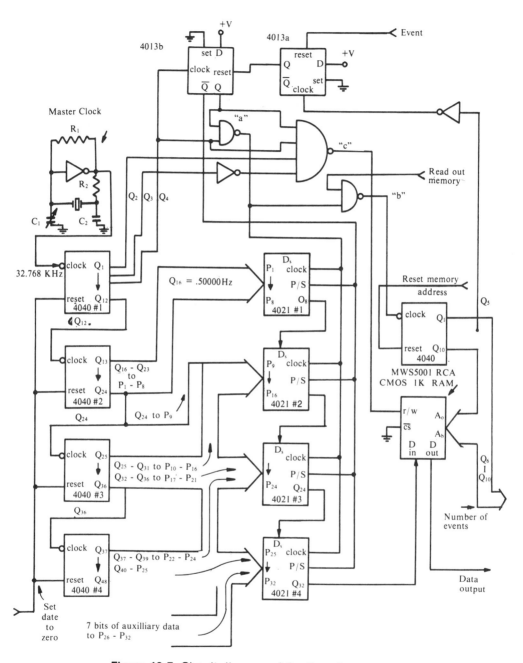

Figure 12-7. Circuit diagram of the Date Recorder.

discuss each section of this design in detail, finishing with a description of the control circuitry and timing diagram.

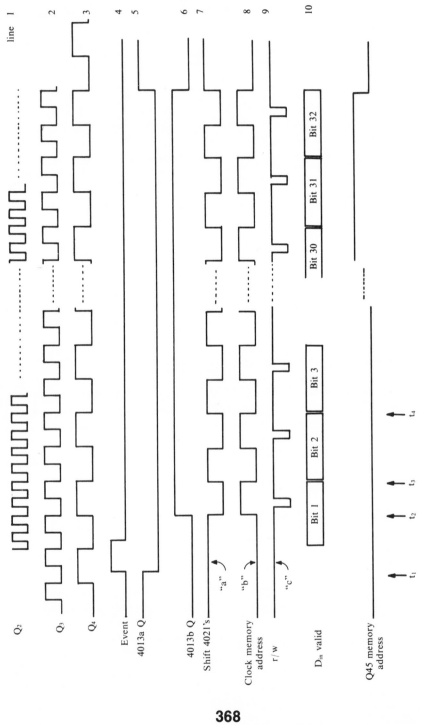

Figure 12-8. State Diagram for Date Recorder.

368

Crystal Oscillator

The oscillation frequency of the master clock should be an integer power of 2 so that a 1 Hz time base can be obtained from the master clock by a succession of "divide-by-two" circuits. The lowest practical master clock frequency should be chosen since this will limit the number of divider stages and minimize power dissipation in the oscillator circuit. Most crystal manufacturers offer a 32.768 kHz (2^{15}) crystal for this type of application. Lower frequency crystals tend to be large and have limited availability and although higher frequency crystals are offered, their only advantage is reduced size.

The oscillator shown in Figure 12-7 uses the crystal π network discussed in Chapter 4, Section 4.5. This circuit needs just one inverter as a gain stage and has the widest tuning range of any crystal oscillator. For this crystal:

$$
\begin{aligned}
R_S &= 5400\,\Omega \\
L_C &= 930.6 \text{ Henries} \\
C_C &= .02537 \text{ pF} \\
C_S &= 15.2 \text{ pF} \\
Q &= 35,000
\end{aligned}
$$

and parallel resonance occurs about 32.768 kHz with a 15 pF parallel load. Since the crystal is specified for nominal operation with a 15 pF load, we set the series combination of C_1 and C_2 (Figure 12-7) equal to 15 pF when the variable capacitor C_1 is at mid-range. Let $C_1 = 5$-60 pF and $C_2 = 30$ pF. When $C_1 = C_2$, the π network has unity inverting gain but when $C_1 = 60$ pF, the net attenuation is about 4. We must choose R_2 so that the total loop gain is greater than 1 under the worst-case conditions. In practice, the loop gain should always be much larger than one since the output of the inverter must be a good square wave. Using Equations 4-9 and 4-10 we find that the lowest π network input resistance (i.e., between C_2 and ground) at resonance is 95K Ω. This occurs when $C_1 = 5$ pF. If we set $R_2 = 100$K Ω we will be assured of a strong signal at the inverter inputs without unduly loading the output of the oscillator. After the circuit is assembled, we adjust C_1 to obtain exactly 32.768 kHz output.

Divider Stages and 25-Bit Counter

In this design, four 12-stage binary counters are connected in series to form a 48-stage binary counter. The first 15 stages are used to derive the 1 Hz clock signal required by the date recorder and the next 25 stages ($Q_{16} - Q_{40}$) are used to store the elapsed time information. The last 8 stages ($Q_{41} - Q_{48}$)

are not used. The CMOS 4040 12-bit counter is an asynchronous design (see Section 9.4.4). Thus, in addition to the external ripple-clocking connection, the internal stages are also ripple-clocked. With 40 total stages a clocking pulse must ripple through 39 stages before reaching Q_{40} for a worst-case transition. The propagation delay from clock pulse in to Q_{12} output for each counter, as specified by the manufacturer, is no more than 8 μs, so a total of $4 \times 8 = 32\mu$s must elapse following a *negative* transition of the master clock before we can guarantee that the counter outputs are valid.

Parallel to Serial Converter

Binary counter outputs Q_{16} - Q_{40} and the 7 bits of auxilliary data are applied to the parallel inputs of a 32-stage parallel-to-serial converter constructed of 4 CMOS 4021 MSI circuits (see Section 9.4.4). Q_{16} is connected to P_1; Q_{17} to P_2 and so on, finishing with Q_{40} which is connected to P_{25}. The 7 auxiliary bits are connected to P_{26} - P_{32} in any fashion. Linked as shown in Figure 12-7, the registers accept parallel data when the P/S line is high irrespective of the serial inputs (D_S) or the clock line. When the P/S goes low, the parallel data at P_1 - P_{32} is locked into the register. Prior to clocking, the contents of the last stage (P_{32} input) are available at the Q_{32} output of the register. With the P/S line low, the stored data advances one stage with each positive clock transition, appearing in succession at Q_{32}. The 7 auxiliary bits are the first to be read out followed by the elapsed time date, MSB first.

CMOS 1024 Bit Memory and Address Register

In the RCA MSW 5001 CMOS 1 k (1024 bits) Random Access Memory (RAM), 1024 separate memory cells are selected, one at a time, by applying a binary number between 0 and 1023 to the address inputs A_0 - A_9.* In Figure 12-7, the addresses are generated by using the first 10 stages of another CMOS 4040 12 stage counter. If the memory address is reset upon field installation then the first event will be written into memory locations 0-31; the second into locations 32-63 and so on with the 32nd event records stored in locations 992-1023. Once the address lines are stable, data at the D_{in} input is written into the selected memory location when the r/w (read/write) line is taken high. The only precaution is that the r/w line must be high for at least 60 ns prior to any address change and remain high for 60 ns following the address change. To ensure proper writing of data into the memory, the r/w line must be low for at least 50 ns prior to a low-to-high

---

* A RAM is a simple extension of the data latch presented in Chapter 8. In a RAM, the total number of data latches is equal to the bit number; 1024 in our example. A specific latch is selected by the address inputs. The read/write input determines whether data is being read out of a selected latch or written into a selected latch.

transition. Data need only be valid at the D_{in} for 30 ns prior to the write command. Although we will be using the memory at substantially reduced speeds, the above requirements must be met. For readout, we simply hold the r/w line high and apply the address of the desired bit. D_{out} will reflect the contents of the selected stage about 150 ns after the address is applied.

If the 4040 memory address register is initially cleared, then its Q_5 output will go low after each series of 32 clock pulses, providing a convenient means of determining when the 32-bit paralled-to-serial register has been fully unloaded into the memory.

Control Circuitry (CC)

We have already outlined *what* the control circuitry must do and now, we will discuss *how* it does it. the CC is best understood in combination with the system timing diagram, Figure 12-8. Following the arrival of an event pulse, system operation proceeds in synchronism with Q_4, an intermediate frequency from the master clock divider circuit. Q_2 and Q_3 of this divider are also brought into the control section, but are only used in conjunction with Q_4 to generate a read/write command for the CMOS RAM. For operation in the record-event mode the "readout memory" line must be high and all 5 4040 counter resets must be low.

Initially, the Q output of the 4013a D type flip-flop is high and the Q output of 4013b is forced low since its reset line is high. With Q-4013b low, the control signals a, b, c, (lines 7, 8, 9 respectively, Figure 12-8) are in the stable states shown prior to t_1. Q - 4013b will be high holding the PS converter in the parallel load mode. Since the 4040's count in a true binary code, stages following Q_4 can be clocked only on the *positive-to-negative* transition of Q_4. As mentioned, about 32 μs are required for the MSB of the 40-stage counter to settle. However, the frequency of Q_4 is only 2048 Hz and so each half cycle will take $\approx$ 224 μs. Therefore, the counter output stages are certain to be stable when the Q_4 undergoes a *negative-to-positive* transition. This is the key to ensuring that valid data is loaded into the P-S register. We now assume that an event pulse arrives at t_1 setting Q-4013a low. With the reset command removed, flip-flop 4013b will load the data present at its D input into the output on the negative to positive transition of the clock Q_4. Thus, at time t_2, following an event pulse, Q-4013b is set low in synchronism with the positive transition of Q_4 latching valid data into the P-S register.

With Q-4013 clocked high, waveforms a, b, c are generated by the logic gates in the control section. As shown, signal "a" will shift the P-S converter on the positive transition and line "b" will clock the memory address on the negative transition; time t_3, t_4 etc. The read/write pulses occur in accordance with the previously discussed RAM requirement. The first read/write pulse occurs before the first clocking of the P-S register or memory clock causing data bit #1 to be entered into the zero memory location. Line 10 of Figure 12-8 illustrates the fact that input data to the RAM is invalid for a brief period following the clocking of the P-S converter. After 32 clock pulses have gone into the P-S converter and memory address counter, Q_5 of the latter device has a positive-to-negative transition which is inverted and applied to the clock of 4013a, returning the entire system to its original state (pre-t_1) when the shift register is fully unloaded. During this entire sequence the elapsed time counters have continued unaffected.

In the reset states, line "a" is left high enabling the NAND gate which generates the "b" control signal. Thus, if we toggle the *readout memory* line, "b" will clock the memory address, providing a simple means of reading out the memory contents without affecting the remainder of the system. As shown in Figure 12-6, memory address outputs Q_6-Q_{10} will have a binary coded record of the total number of events which have occurred, assuming that this counter is initially reset upon field installation.

12.4 CONCLUSION

A first approximation to the solution of both analog and digital system design problems is to draw a detailed block diagram. In the case of analog designs, reduction of the individual blocks to hardware proceeds in a fairly regular fashion. To complete a digital design, we need a detailed timing diagram in addition to the block diagram. Since the blocks of a complex digital design are highly interrelated, reduction of the blocks to hardware form can frequently be quite difficult.

In any design, it is easy to construct a block diagram which is unrealistic or impractical; contrast for example our first shot at the data recorder block diagram Figure 12-5 with the final version. This difficulty can only be minimized through the application of common sense, experience and a very good knowledge of the types of hardware available.

As a last note, much ado is usually made about reducing parts count. Whereas this is certainly a fine design goal, the designer may stifle his creativity by *approaching* his design from this angle. Usually, it is best to proceed "carte blanche" and get a working, though perhaps overly complex, circuit design and detailed timing diagram down on paper. *Then,* begin the process of simplification by combining circuit functions where possible and eliminating redundant and/or unnecessary circuit functions.

INDEX

INDEX

375

E